Technology in a Globalizing World

snova
New York

Technology in a Globalizing World

Nonlinear Systems: Chaos, Advanced Control and Application Perspectives
Piyush Pratap Singh, PhD (Editor)
2022. ISBN: 978-1-68507-660-3 (Softcover)
2022. ISBN: 979-8-88697-001-2 (eBook)

Multidisciplinary Science and Advanced Technologies
Fernando Gomes, PhD (Editor)
Kaushik Pal, PhD (Editor)
Thinakaran Narayanan (Editor)
2021. ISBN: 978-1-53618-959-9 (Hardcover)
2021. ISBN: 978-1-53619-198-1 (eBook)

Issues with Facial Recognition Technology
Warren Lambert (Editor)
2020. ISBN: 978-1-53618-973-5 (Hardcover)
2020. ISBN: 978-1-53619-000-7 (eBook)

The Pulp and Paper Industry: Production, Management and Technology
Mette P. Kristensen (Editor)
2020. ISBN: 978-1-53617-919-4 (eBook)

A Closer Look at Educational Technology
Maria A. Clausen (Editor)
2019. ISBN: 978-1-53616-215-8 (Softcover)
2019. ISBN: 978-1-53616-216-5 (eBook)

More information about this series can be found at
https://novapublishers.com/product-category/series/technology-in-a-globalizing-world/

Mari F. Burke
Editor

Facial Recognition Technology

Usage by Federal Law Enforcement

snova
New York

Library of Congress Cataloging-in-Publication Data

ISBN: 979-8-88697-124-8

Published by Nova Science Publishers, Inc. † New York

Contents

Preface

Facial recognition—a type of biometric technology—mimics how people identify or verify others by examining their faces. Recent advancements have increased the accuracy of automated FRT resulting in increased use across a range of applications. As the use of facial recognition technology (FRT) has continues to expand, it has become increasingly important to understand its use across the federal government in a comprehensive way. Law enforcement agencies' use of FRT has received increased attention from policymakers and the public. Some of the concerns raised revolve around the accuracy of the technology, including potential race-, gender-, and age-related biases; the process of collecting, retaining, and securing images contained in various facial recognition databases; public notification of the use of facial recognition and other image-capturing technology; and policies or standards governing law enforcement agencies' use of the technology. Some of these concerns have manifested in actions such as federal, state, and city efforts to prohibit or bound law enforcement agencies' use of FRT. In addition, some companies producing facial recognition software have placed new barriers to law enforcement using their technologies. This book provides an overview of federal law enforcement agencies' use of FRT, including the current status of scientific standards for its use.

Chapter 1 - Law enforcement agencies' use of facial recognition technology (FRT), while not a new practice, has received increased attention from policymakers and the public. Some of the concerns raised revolve around the accuracy of the technology, including potential race-, gender-, and age-related biases; the process of collecting, retaining, and securing images contained in various facial recognition databases; public notification of the use of facial recognition and other image-capturing technology; and policies or standards governing law enforcement agencies' use of the technology. Some of these concerns have manifested in actions such as federal, state, and city efforts to prohibit or bound law enforcement agencies' use of FRT. In addition, some companies producing facial recognition software have placed new barriers to law enforcement using their technologies. FRT is one of several biometric technologies employed by law enforcement agencies, which also include fingerprint, palm print, DNA, and iris scans. FRT can be used by law enforcement for a variety of purposes such as generating investigative leads, identifying victims of crimes, helping sort faces in photos that are part of forensic evidence, and helping verify the identity of inmates before they are released from prison. However, the frequency and extent to which FRT is used at various phases of the criminal justice system is unknown. It is most often discussed by law enforcement officials as being used to help identify suspects.

The Federal Bureau of Investigation (FBI) is a leading federal law enforcement agency in the use of FRT. The bureau operates two programs that support the use of the technology: (1) the Next Generation Identification–Interstate Photo System (NGI-IPS), largely supporting state and local law enforcement; and (2) the Facial Analysis, Comparison, and Evaluation (FACE) Services Unit, supporting FBI investigations. NGI-IPS contains criminal mugshots, and the system allows authorized law enforcement users (primarily state and local) to search the database for potential investigative leads. The FACE Services Unit supports FBI investigations by searching probe photos of unknown persons against faces in NGI-IPS and other federal and state facial recognition systems authorized for FBI use. A facial recognition search alone does not provide law enforcement with a positive identification; the results need to be manually reviewed and compared by an officer trained in facial comparison. Further, law enforcement agencies are prohibited from relying solely on the results of a search in NGI-IPS to take a law enforcement action (e.g., making an arrest). The FBI maintains an NGI Policy and Implementation Guide that outlines policies surrounding use of NGI-IPS. Authorized law enforcement users of NGI-IPS are required to follow these policies as well as established standards for performing facial comparison. While FRT is generally used by law enforcement agencies to help generate potential investigative leads, it is also employed by U.S. border enforcement officials to assist with verifying travelers' identities. The Department of Homeland Security (DHS) and Customs and Border Protection (CBP) use the Traveler Verification Service (TVS). TVS compares the travelers' live photographs (taken, for example, by a gate agent) to a gallery of photographs and provides a match or no match result; a no match results in a manual check by an official to verify a traveler's identity.

Guidelines and recommendations regarding law enforcement use of FRT have been produced by the Facial Identification Scientific Working Group (FISWG). FISWG is one of the various scientific working groups that support the Organization of Scientific Area Committees for Forensic Science (administered by the National Institute of Standards and Technology, which facilitates standards development, including for FRT). FISWG has published a number of FRT-related documents for forensic science practitioners. For instance, they have guidelines and recommendations for establishing and conducting training on facial comparison, guides for capturing facial images that can be used in facial recognition systems, and recommended methods and techniques for using facial recognition systems. As policymakers consider legislation and oversight on law enforcement agencies' use of FRT, they may evaluate how the accuracy of these systems is defined and assessed by law enforcement; how to support or restrict the technology's use by federal, state, and local law enforcement—and potential implications; and how to balance privacy and security concerns with supporting lawful criminal justice activities.

Chapter 2 - Facial recognition—a type of biometric technology—mimics how people identify or verify others by examining their faces. Recent advancements have increased the accuracy of automated FRT resulting in increased use across a range of applications. As the use of FRT continues to expand, it has become increasingly important to understand its use across the federal government in a comprehensive way. GAO was asked to review the extent of FRT use across the federal government. This chapter identifies and describes (1) how agencies used FRT in fiscal year 2020, including any related research and development and interactions with non-federal entities, and (2) how agencies plan to expand their use of FRT through fiscal year 2023. GAO surveyed the 24 agencies of the Chief Financial Officers Act of 1990, as amended, regarding their use of facial recognition technology. GAO also interviewed agency officials and reviewed documents, such as system descriptions, and information provided by agencies that reported using the technology.

Chapter 3 - Federal agencies that employ law enforcement officers use facial recognition technology to assist criminal investigations, among other activities. For example, the technology can help identify an unknown individual in a photo or video surveillance. This statement describes (1) the ownership and use of facial recognition technology by federal agencies that employ law enforcement officers, (2) the types of activities these agencies use the technology to support, and (3) the extent that these agencies track employee use of facial recognition technology owned by non-federal entities, including the potential privacy and accuracy implications. This statement is based on GAO's June 2021 report on federal law enforcement's use of facial recognition technology (GAO-21-518). To conduct that prior work, GAO administered a survey questionnaire to 42 federal agencies that employ law enforcement officers regarding their use of the technology. GAO also reviewed relevant documents and interviewed agency officials. The June 2021 report was a public version of a sensitive report that GAO issued in April 2021. Information that agencies deemed sensitive was omitted from the June 2021 report and this statement.

Chapter 4 - Federal agencies that employ law enforcement officers can use facial recognition technology to assist criminal investigations, among other activities. For example, the technology can help identify an unknown individual in a photo or video surveillance. GAO was asked to review federal law enforcement use of facial recognition technology. This chapter examines the 1) ownership and use of facial recognition technology by federal agencies that employ law enforcement officers, 2) types of activities these agencies use the technology to support, and 3) the extent that these agencies track employee use of facial recognition technology owned by non-federal entities. GAO administered a survey questionnaire to 42 federal agencies that employ law enforcement officers regarding their use of the technology. GAO also reviewed documents (e.g., system descriptions) and interviewed officials from selected agencies (e.g., agencies that owned facial recognition technology). This is a public version of a sensitive report that GAO issued in April 2021. Information that agencies deemed sensitive has been omitted.

Chapter 5 - Facial recognition technology (FRT), which uses algorithms to compare faces against a collection of known faces, is a biometric tool employed by law enforcement, including U.S. Customs and Border Protection (CBP) officials. FRT can be used to verify the identity of travelers and certain noncitizens awaiting entry.

FRT is used in two primary ways:

- identification searches, comparing features of a probe photo with those in a database of images to produce a gallery of potential matches ranked by similarity (not a single affirmative match), and
- identity verifications, comparing a probe photo against known photos to produce a binary, match/no match result.

Chapter 6 - On January 6, 2021, U.S. Capitol security was violently breached while Congress was in session to vote on the certification of the 2020 presidential election. Many participants in the attack reportedly intended to disrupt this process. Some clashed with law enforcement, leaving five dead and many more injured. Some have been charged with violations including crimes against persons and federal property. Some allegedly carried firearms and other weapons. Law enforcement also uncovered explosive devices near Capitol grounds. Participants, media, and others documented the attack on the Capitol in photographs and video. Law enforcement, such as the Federal Bureau of Investigation (FBI), is seeking this digital content and other evidence to assist its investigation. Among the tools that law enforcement agencies may employ to identify potential criminal suspects depicted in digital content is facial recognition technology (FRT). While some have reported a possible uptick in law enforcement use of FRT following the Capitol attack, the extent to which FRT has been used to identify suspected rioters is unknown.

Chapter 1

Federal Law Enforcement Use of Facial Recognition Technology*

Kristin Finklea, Laurie A. Harris, Abigail F. Kolker and John F. Sargent Jr.

Summary

Law enforcement agencies' use of facial recognition technology (FRT), while not a new practice, has received increased attention from policymakers and the public. Some of the concerns raised revolve around the accuracy of the technology, including potential race-, gender-, and age-related biases; the process of collecting, retaining, and securing images contained in various facial recognition databases; public notification of the use of facial recognition and other image-capturing technology; and policies or standards governing law enforcement agencies' use of the technology. Some of these concerns have manifested in actions such as federal, state, and city efforts to prohibit or bound law enforcement agencies' use of FRT. In addition, some companies producing facial recognition software have placed new barriers to law enforcement using their technologies.

FRT is one of several biometric technologies employed by law enforcement agencies, which also include fingerprint, palm print, DNA, and iris scans. FRT can be used by law enforcement for a variety of purposes such as generating investigative leads, identifying victims of crimes, helping sort faces in photos that are part of forensic evidence, and helping verify the identity of inmates before they are released from

* This is an edited, reformatted and augmented version of the Congressional Research Service Publication No. R46586, dated October 27, 2020.

In: Facial Recognition Technology
Editor: Mari F. Burke
ISBN: 979-8-88697-124-8

prison. However, the frequency and extent to which FRT is used at various phases of the criminal justice system is unknown. It is most often discussed by law enforcement officials as being used to help identify suspects.

The Federal Bureau of Investigation (FBI) is a leading federal law enforcement agency in the use of FRT. The bureau operates two programs that support the use of the technology: (1) the Next Generation Identification– Interstate Photo System (NGI-IPS), largely supporting state and local law enforcement; and (2) the Facial Analysis, Comparison, and Evaluation (FACE) Services Unit, supporting FBI investigations. NGI-IPS contains criminal mugshots, and the system allows authorized law enforcement users (primarily state and local) to search the database for potential investigative leads. The FACE Services Unit supports FBI investigations by searching *probe* photos of unknown persons against faces in NGI-IPS and other federal and state facial recognition systems authorized for FBI use. A facial recognition search alone does not provide law enforcement with a positive identification; the results need to be manually reviewed and compared by an officer trained in facial comparison. Further, law enforcement agencies are prohibited from relying solely on the results of a search in NGI-IPS to take a law enforcement action (e.g., making an arrest). The FBI maintains an NGI Policy and Implementation Guide that outlines policies surrounding use of NGI-IPS. Authorized law enforcement users of NGI-IPS are required to follow these policies as well as established standards for performing facial comparison.

While FRT is generally used by law enforcement agencies to help generate potential investigative leads, it is also employed by U.S. border enforcement officials to assist with verifying travelers' identities. The Department of Homeland Security (DHS) and Customs and Border Protection (CBP) use the Traveler Verification Service (TVS). TVS compares the travelers' *live photographs* (taken, for example, by a gate agent) to a gallery of photographs and provides a *match* or *no match* result; a no match results in a manual check by an official to verify a traveler's identity.

Guidelines and recommendations regarding law enforcement use of FRT have been produced by the Facial Identification Scientific Working Group (FISWG). FISWG is one of the various scientific working groups that support the Organization of Scientific Area Committees for Forensic Science (administered by the National Institute of Standards and Technology, which facilitates standards development, including for FRT). FISWG has published a number of FRT-related documents for forensic science practitioners. For instance, they have guidelines and recommendations for establishing and conducting training on facial comparison, guides for capturing facial images that can be used in facial recognition systems, and recommended methods and techniques for using facial recognition systems.

As policymakers consider legislation and oversight on law enforcement agencies' use of FRT, they may evaluate how the accuracy of these systems is defined and assessed by law enforcement; how to support or restrict the technology's use by federal, state, and local law enforcement—and potential implications; and how to balance privacy and security concerns with supporting lawful criminal justice activities.

Law enforcement agencies' use of facial recognition technology (FRT), while not a new practice, has received increased attention from policymakers and the public. In the course of carrying out their duties, federal law enforcement agencies may use FRT for a variety of purposes. For instance, the Federal Bureau of Investigation (FBI) uses the technology to aid its investigations, and the bureau provides facial recognition assistance to federal, state, local, and tribal law enforcement partners. State, local, and tribal law enforcement have also adopted facial recognition software systems to assist in various phases of investigations. In addition, border officials use facial recognition for identity verification purposes.

The use of FRT by law enforcement agencies has spurred questions on a range of topics. Some primary concerns revolve around the accuracy of the technology, including potential race-, gender-, and age-related biases; the collection, retention, and security of images contained in various facial recognition databases; public notification regarding the use of facial recognition and other image capturing technology; and policies or standards governing law enforcement agencies' use of the technology. Some of these concerns have manifested in actions such as federal, state, and city efforts to prohibit or bound law enforcement agencies' use of FRT.[1] In addition, some companies producing facial recognition software, such as Microsoft, IBM, and Amazon, have enacted new barriers to law enforcement using their technologies.[2]

This chapter provides an overview of federal law enforcement agencies' use of FRT, including the current status of scientific standards for its use. The chapter includes a discussion of how FRT may be used by law enforcement agencies with traditional policing missions as well as by those charged with

---

[1] See, for example, Dave Lee, "San Francisco is First US City to Ban Facial Recognition," *BBC*, May 15, 2019; and Dustin Gardiner, "California Blocks Police From Using Facial Recognition in Body Cameras," *San Francisco Chronicle*, October 8, 2019.

[2] See, for example, Jay Greene, "Microsoft Won't Sell Police Its Facial-Recognition Technology, Following Similar Moves by Amazon and IBM," *The Washington Post*, June 11, 2020.

securing the U.S. borders. It also discusses considerations for policymakers debating whether or how to influence federal, state, and local law enforcement agencies' use of FRT.

Conceptualizing Facial Recognition Technology

The term *facial recognition technology* can have different meanings for law enforcement agencies, policymakers, and the public, and the process of using facial recognition in a law enforcement context can involve various technologies and actors. Broadly, as technology experts have noted, "[t]here is no one standard system design for facial recognition systems. Not only do organizations build their systems differently, and for different environments, but they also use different terms to describe how their systems work."[3] The following key terms are provided to help in understanding facial recognition technologies and processes in this chapter.[4]

Face detection technology determines whether a digital image contains a face.

Facial classification algorithms analyze a face image to produce an estimate of age, sex, or some other property, but do not identify the individual. An example application of this would be retail stores using facial classification to gather data on the gender and age ranges of people visiting a store, without identifying each shopper individually.

Facial comparison and facial identification are often used in the same context. They involve a human manually examining the differences and similarities between facial images, or betweena live subject and facial images, for the purpose of determining if they represent the sameperson.

Facial comparison has three broad categories: assessment, review, and examination. Facial assessment is a quick image-to-image or image-to-person comparison, typically carried out in screening or access control situations and is the least rigorous form of facial comparison. Facial review (often used in

---

[3] Partnership on AI, *Understanding Facial Recognition Systems*, February 19, 2020, p. 3, https://www.partnershiponai.org/wp-content/uploads/2020/02/Understanding-Facial-Recognition-Paper_final.pdf.

[4] These terms are taken or adapted from the Facial Identification Scientific Working Group (FISWG), *FISWG Glossary Version 2.0*, October 25, 2019; FISWG, *Facial Comparison Overview and Methodology Guidelines*, October 25, 2019; and National Institute of Standards and Technology (NIST), *Face Recognition Vendor Test (FRVT), Part 3: Demographic Effects (NIST IR8280)*, December 19, 2019, https://doi.org/10.6028/ NIST.IR.8280. Some editorial changes have been made.

investigative, operational, or intelligence gathering applications) and facial examination (often used in a forensic applications) are increasingly rigorous levels of image comparison and should involve verification by an additional reviewer or examiner. They may involve a formal, systematic examination of facial images.

Facial recognition broadly involves the automated searching of a facial image (a probe) against a known collection or database of photos.

Facial recognition algorithms compare identity information from facial features in two face image samples and produce a measure of similarity (sometimes called a match score) between them; this information can be used to determine whether the same person is in both images.

Images that have a similarity score above a defined threshold are presented to the user. There are two ways in which facial recognition algorithms work to compare images:

- One-to-one verification algorithms compare a photo of someone claiming a specific identity with a stored image(s) of that known identity to determine if it is the same person. Uses of these algorithms can include unlocking a smartphone and verifying identities at a security checkpoint.
- One-to-many identification search algorithms compare features of a probe photo with all those in a gallery of images. The algorithms can provide either a fixed number of the most similar candidates, or all candidates with a similarity score above a preset threshold, for human review.[5] These algorithms may be used for purposes such as identifying potential suspect leads from a mugshot database.

Probe refers to the facial image or template searched against a gallery or database of photos in a facial recognition system.

Real-time facial recognition involves facial recognition algorithms that can be used while a video recording is taking place in order to determine in real time whether an individual in a video matches with a list of candidates in a database of photos.

Threshold refers to any real number against which similarity scores are compared to produce a verification decision or gallery of images.

---

[5] This process is described in more detail in the "Accuracy and Interpretation of Results" section of this report.

Scientific Standards and Facial Recognition Technology

NIST's Role in Facial Recognition Technology

The National Institute of Standards and Technology (NIST) is a non-regulatory federal agency within the Department of Commerce charged with promoting U.S. innovation and industrial competitiveness by advancing measurement science, standards, and technology. Among its key roles, NIST is the lead federal agency for metrology (the science of measurement) and facilitates standards development, two key elements in the development and deployment of FRT.[6]

NIST's work in FRT includes the following:

- *Research* to improve the accuracy, quality, usability, interoperability, and consistency of FRT identity management systems;
- *Testing and Evaluation* to provide tools and support for evaluating the effectiveness of FRT prototypes and products;
- *Technical Guidance and Support* to assist federal law enforcement and other federal government agencies in the use of FRT; and
- *Standards* to facilitate the development of scientifically valid, fit-for-purpose FRT standards and to ensure that U.S. interests are represented in international arenas.

NIST collaborates with other federal agencies, law enforcement agencies, industry, and academic partners in these and related activities. Detailed information on NIST efforts in research, testing and evaluation, technical guidance and support, and standards development related to FRT are included in the Appendix.

Facial Identification Scientific Working Group

NIST administers the Organization of Scientific Area Committees (OSAC) for Forensic Science, which is a collaboration of more than 550 forensic scientific practitioners and experts across government, academia, and industry.[7] The OSAC for Forensic Science works to facilitate "the development of technically sound, science-based standards through a formal standard

[6] Facial recognition technology is a subfield of biometrics. Biometrics is the measurement and analysis of unique physical or behavioral characteristics.

[7] For more information, see https://www.nist.gov/topics/organization-scientific-area-committees-forensic-science.

developing organization (SDO) process."[8] It also evaluates existing standards developed by SDOs and may place them on the OSAC Registry,[9] which contains approved scientifically sound forensic science standards for specific disciplines, such as facial identification, digital evidence, or bloodstain pattern analysis. The OSAC also promotes the use of these OSAC Registry-approved standards by the forensic science community.

The OSAC for Forensic Science is supported by a number of scientific working groups, which are collaborations between forensic science laboratories and practitioners "to improve discipline practices and build consensus standards."[10] There are over 20 scientific working groups across a range of disciplines including facial identification, DNA analysis, and latent fingerprint examination. The mission of the Facial Identification Scientific Working Group (FISWG) "is to develop consensus standards, guidelines and best practices for the discipline of image-based comparisons of human features, primarily face, as well as to provide recommendations for research and development activities necessary to advance the state of the science in this field."[11]

FISWG has published a number of guidelines and recommendations for forensic science practitioners. For instance, they have guidelines and recommendations for conducting and establishing training on facial comparison, guides for capturing facial images that can be used in facial recognition systems, and recommended methods and techniques for using facial recognition systems.[12]

How FRT May be Used by Federal Law Enforcement Agencies

Law enforcement agencies' use of FRT has received attention from policymakers and the public over the past several years. There have been heightened concerns following several revelations, including that Clearview AI, a company that developed image-search technology used by law

---

[8] Ibid.

[9] See https://www.nist.gov/topics/organization-scientific-area-committees-forensic-science/osac-registry.

[10] See https://fiswg.org/about_swgs.html.

[11] See https://fiswg.org/objectives.html.

[12] See https://fiswg.org/documents.html.

enforcement agencies around the country, had amassed a database of over 3 billion images against which probe photos could be compared.[13]

FRT is one of several biometric technologies employed by law enforcement agencies, which also include fingerprint, palm print, DNA and iris scans. FRT can be used by law enforcement for a variety of purposes such as generating investigative leads, identifying victims of crimes, facilitating the examination of forensic evidence, and helping verify the identity of individuals being released from prison.[14] Press releases and statements from the Department of Justice (DOJ) highlight how the technology has been used in the criminal justice system.

- FRT has been used to help generate suspect leads. In one case, FBI agents used the technology, via the Mississippi Fusion Center, to identify a potential suspect in an interstate stalking case who had allegedly been harassing high school girls through their Twitter accounts.[15] The suspect was later sentenced to 46 months imprisonment and three years of supervised release for this stalking.[16]
- FRT may also be used to help identify victims. For example, officials have noted FRT was used to help identify "an accident victim lying unconscious on the side of the road."[17]
- FRT, along with other pieces of evidence, has been used to support probable cause in affidavits in support of criminal complaints. In one case, an FBI agent cited the use of FRT in a criminal complaint against a bank robbery suspect. The agent noted that images from the bank's surveillance footage were run against facial recognition software, and a photo of the suspect was returned as a possible match. Investigators then interviewed associates of the suspect, who identified him as the man in the bank surveillance footage.[18]

---

[13] Kashmir Hill, "The Secretive Company That Might End Privacy as We Know It," *The New York Times*, February 10, 2020.

[14] Remarks by an FBI representative at the President's Commission on Law Enforcement and the Administration of Justice, April 21, 2020.

[15] U.S. Attorney's Office, Southern District of Indiana, "Mississippi man faces interstate stalking charges for five-year- long crime against Evansville area high schoolers," press release, June 28, 2018.

[16] U.S. Attorney's Office, Southern District of Indiana, "Mississippi man sentenced for internet stalking young Evansville women," press release, August 5, 2019.

[17] Remarks by an FBI representative at the President's Commission on Law Enforcement and the Administration of Justice, April 21, 2020.

[18] United States of America v. Terrance Maurice Goss, U.S. District Court for the Middle District of Florida, Criminal Complaint, January 18, 2019.

Notably, the frequency and extent to which FRT is used at various phases of the criminal justice system (from generating leads and helping establish probable cause for an arrest or indictment, to serving as evidence in courtrooms) is unknown.[19] It is most often discussed as being employed during investigations by law enforcement officials. Of note, FRT is generally used by law enforcement in one-to-many searches to produce a gallery of potential suspects ranked by similarity and not to provide a single affirmative match. As such, the technology currently might not be relied upon in the same way that other biometric evidence might. Rather, it is the results of an investigator's facial review between a probe face and the gallery of images produced from running a probe face through facial recognition software that might be used as evidence contributing to an arrest and prosecution.

While FRT is used by a number of federal law enforcement agencies, the next section of this chapter highlights the FBI's use of it, as the bureau has a leading role in federal law enforcement's employment of the technology.

FBI Use of FRT

The FBI's Criminal Justice Information Services (CJIS) Division operates two programs that support the FBI's use of FRT: (1) the Next Generation Identification–Interstate Photo System (NGI-IPS), largely supporting state and local law enforcement; and (2) the Facial Analysis, Comparison, and Evaluation (FACE) Services Unit, supporting FBI investigations.

Next Generation Identification–Interstate Photo System (NGI-IPS)

NGI-IPS contains criminal mugshots that have associated 10-print fingerprints and criminal history records. This system allows authorized federal, state, local, and tribal law enforcement users to search the database for potential investigative leads. To use NGI-IPS,

> [a] law enforcement agency submits a "probe" photo that is obtained pursuant to an authorized law enforcement investigation, to be searched against the mugshot repository. The NGI-IPS returns a gallery of "candidate" photos of 2-50 individuals (the default is 20). During the

---

[19] In at least one case, in California, a judge allowed a suspect's criminal defense team to introduce evidence from biometric facial recognition technology. For more information, see "A First: Biometrics Used to Sentence Criminal," *Homeland Security Newswire*, February 1, 2011.

second step of the process, the law enforcement agencies then manually review the candidate photos and perform further investigation to determine if any of the candidate photos are the same person as the probe photo.[20]

The FBI notes that a facial recognition search in NGI-IPS cannot alone provide a positive identification; the results need to be manually reviewed by a trained officer.[21] Further, law enforcement agencies that submit a probe photo for a search in NGI-IPS are prohibited from relying solely on the results of this search to take a formal law enforcement action (e.g., making an arrest).[22]

Photos in NGI-IPS

NGI-IPS "contains over 93 million civil photos, criminal photos, and scars, marks and tattoo images. Of this number, over 38 million criminal photos are available for facial recognition searching by law enforcement agencies."[23] The photos in NGI-IPS are separated into various groups: a criminal identity group (mugshots pursuant to arrest),[24] a civil identity group (those submitted for criminal background checks for non-criminal justice purposes such as employment and security clearances), and an unsolved photo file (UPF, which contains photos of unknown subjects reasonably suspected of a felony crime against a person). The criminal identity group is automatically available for facial recognition searching. If an individual has photos associated with both the criminal and civil identity groups, all photos become associated with the criminal identity group and are available for searching. A law enforcement agency submitting a probe photo to NGI-IPS for searching can affirmatively request to search against the UPF in addition to the criminal identity group.

Facial Analysis, Comparison, and Evaluation (FACE) Services Unit

While NGI-IPS primarily supports state, local, and tribal law enforcement partners, the FACE Services Unit supports FBI investigations. Specifically,

---

[20] FBI testimony before U.S. Congress, House Committee on Oversight and Reform, *Facial Recognition Technology (Part II): Ensuring Transparency in Government Use*, 116th Cong., 1st sess., June 4, 2020.

[21] As noted elsewhere, authorized users of NGI-IPS must receive training in the use of facial recognition technology.

[22] See https://www.fbibiospecs.cjis.gov/Face.

[23] FBI, *Privacy Impact Assessment for the Next Generation Identification – Interstate Photo System*, October 29, 2019.

[24] The mugshots taken pursuant to arrest are not indicative of actual criminality or guilt.

FACE Services supports FBI field offices, operational divisions, and legal attachés (and sometimes federal partners) on open investigations and, in limited circumstances, on closed cases. The FACE Services Unit searches "probe photos that have been collected pursuant to the Attorney General guidelines as part of an authorized FBI investigation, and they are not retained."[25] These probe photos are searched against faces in NGI- IPS as well as other federal and state facial recognition systems authorized for FBI use. The FACE Services Unit then, through facial review, compares the probe photo against the candidate gallery of faces produced from the search to help identify potential investigative leads.

FRT used by law enforcement agencies at the border

While FRT is generally used by law enforcement agencies to help generate potential investigative leads, it is also employed by U.S. border enforcement officials to assist with verifying travelers' identities. The Department of Homeland Security (DHS) is developing an automated biometric entry-exit system for foreign nationals traveling into and out of the country.[26] The entry system has been implemented,[27] but the exit system has yet to be fully operationalized.[28] In developing the exit system, DHS and Customs and Border Protection (CBP) have piloted various programs using an array of biometric technologies (e.g., fingerprints, iris scans, and facial recognition) and determined that facial recognition was the optimal approach because of the speed with which it could be used and its relative accuracy. The FRT program they have implemented is the Traveler Verification Service (TVS).[29]

---

[25] FBI testimony before U.S. Congress, House Committee on Oversight and Reform, *Facial Recognition Technology (Part II): Ensuring Transparency in Government Use*, 116th Cong., 1st sess., June 4, 2020. Department of Justice policies and procedures—including guidelines for investigations—are outlined in the *Justice Manual*, available at https://www.justice.gov/jm/justice-manual.

[26] For more information, see CRS In Focus IF11634, *Biometric Entry-Exit System: Legislative History and Status.*

[27] The entry system, fully implemented in December 2006, utilizes biometrics such as fingerprints and digital photographs.

[28] There have been "various longstanding planning, infrastructure, and staffing challenges" to developing and implementing the biometric exit system, including airports' lack of secure inspection areas for outbound travelers. See U.S. Government Accountability Office, *DHS Has Made Progress in Planning for a Biometric Air Exit System and Reporting Overstays, but Challenges Remain*, GAO-17-170, February 17, 2017.

[29] Additional information, including privacy documents, is available at https://www.dhs.gov/publication/dhscbppia- 056-traveler-verification-service.

TVS is a public-private partnership between the federal government and private airlines, airports, and cruise lines. It is deployed by CBP and the Transportation Security Administration (TSA). TVS currently operates in 27 airports, 7 seaports, and 5 border locations across the United States, as well as 4 international preclearance locations. TVS currently captures roughly 60% of *in-scope* travelers (i.e., foreign nationals aged 14-79) departing the United States via commercial air carriers and 20% of in-scope arriving travelers.[30] CBP's goal is to capture 97% of all in-scope departing commercial air travelers by 2022.[31] TVS compares the travelers' *live photographs* (taken, for example, by a gate agent) to a gallery of photographs. The composition of the galleries depends on the travel context. For air and sea travelers, CBP uses biographic data obtained from flight and ship manifests via the Advance Passenger Information System[32] to gather all associated facial images from DHS holdings (e.g., photographs from U.S. passports, U.S. visas, CBP entry inspections, and any other DHS encounters). For pedestrians and vehicle travelers, the gallery consists of photographs of frequent crossers at that port of entry.

TVS provides a *match* or *no match* result within two seconds.[33] In case of the latter result, the traveler's identity is checked manually by a CBP agent.

Federal Law Enforcement FRT Policy Guidance

The FBI maintains an NGI Policy and Implementation Guide that outlines policies surrounding use of NGI-IPS. Authorized law enforcement users of NGI-IPS are required to follow these policies as well as FISWG standards for performing facial comparison.[34] Policies outlined in the guide include information on how to

---

[30] Based on CRS discussions with CBP officials on January 30, 2020, and CRS email communication with CBP officials on August 12, 2020.

[31] Department of Homeland Security (DHS), *Transportation Security Administration and U.S. Customs and Border Protection: Deployment of Biometric Technologies,* Report to Congress, August 30, 2019, p. 5.

[32] APIS collects biographic data such as gender, date of birth, travel document type and number, and nationality, Ibid. p. 30.

[33] U.S. Customs and Border Control, *Traveler Verification Service for Simplified Travel,* August 2018.

[34] FISWG has Guidelines and Recommendations for Facial Comparison Training to Competency, available at https://fiswg.org/FISWG_Training_Guidelines_Recommendations_v1.1_2010_11_18.pdf.

- submit photos for enrollment in NGI-IPS,
- conduct an investigative photo search,
- retrieve additional biometrics associated with a probable candidate generated from a search of NGI-IPS,
- notify the FBI of a potential match resulting from an investigative photo search,
- request an audit trail for a biometric set that an authorized user enrolled in NGI- IPS, and
- delete a biometric set that an authorized user enrolled in NGI-IPS.[35]

The FBI outlines technical requirements for using NGI-IPS in an Electronic Biometric Transmission Specification document that it provides to system users. In addition, it asks that users of NGI-IPS use its Mugshot Implementation Guide as a reference for submitting proper facial images to the FBI. The guide notes that image quality is affected by the camera, background, lighting, and subject posing.[36]

The FBI requires that users of NGI-IPS complete facial recognition training that meets FISWG guidelines. To facilitate this requirement, the FBI provides facial comparison and identification training, which "is designed to provide the skills and knowledge to professionals from the law enforcement and intelligence communities working in the fields of face recognition and face comparison. It also provides students with awareness and understanding of the face comparison discipline. This training is consistent with the guidelines and recommendations outlined by [FISWG]."[37] FISWG notes that "[t]he level of training necessary to conduct facial comparison is dependent upon the source, quality, quantity, and complexity of the images that are being analyzed and the purpose of the analysis."[38] As outlined by FISWG, basic level training for facial comparison includes, among other things, an understanding of the principles of facial comparison, including

- assessing the quality of a facial image to determine the value for examination;

---

35 FBI, Next Generation Identification (NGI) Interstate Photo System (IPS) Policy and Implementation Guide: Version 1.3, April 23, 2015.

36 FBI, Mugshot Implementation Guide: Photographic Considerations Related to Facial Recognition Software and Booking Station Mug Shots, April 25, 2013.

37 FBI, *Biometric and Criminal History Record Training*, https://www.fbi.gov/services/cjis/fingerprints-and-other-biometrics/biometric-and-criminal-history-record-training.

38 FISWG, Guide for Role-Based Training in Facial Comparison, July 17, 2020, pp. 1-2.

- using a process of "Analysis, Comparison, Evaluation, and Verification (ACE- V)";
- understanding the methods of comparison, such as one-to-one facial examination;
- understanding the levels of conclusion;
- having the ability to render proper conclusions;
- understanding the concept and effects of cognitive bias, including confirmation bias; and
- understanding the benefits of verification by another qualified reviewer or examiner.[39]

The FBI has also conducted audits to evaluate whether users of its facial recognition systems are in compliance with the policies surrounding their use. In congressional testimony, the FBI indicated that as of May 2019, nine FBI audits of NGI-IPS revealed no findings of non-compliance and no observations of unauthorized requests or misuse of NGI-IPS; in addition, a 2018 FBI audit of the FACE Services Unit indicated that the unit is operating in accordance with FBI policies and relevant privacy laws.[40]

Policy Considerations Surrounding Federal Law Enforcement Use of FRT

Accuracy and Interpretation of Results

The accuracy of FRT has come under scrutiny, independent of law enforcement's use of the technology including an officer's review of potential matches. When considering accuracy, there are a number of possible outcomes in both one-to-many identification searches (such as those used by the FBI's NGI-IPS) and one-to-one verifications (such as those used by the CBP's TVS).

Facial recognition systems may return an accurate match (i.e., a true positive result, or hit),[41] an accurate non-match (i.e., a true negative, or correct

---

[39] FISWG, Guidelines and Recommendations for Facial Comparison Training to Competency, November 18, 2010, pp. 1-2.

[40] Statement for the Record of Kimberly Del Greco, Federal Bureau of Investigation, before U.S. Congress, House Committee on Oversight and Reform, Facial Recognition Technology: Ensuring Transparency in Government Use, 116th Cong., 1st sess., June 4, 2019.

[41] In a one-to-many identification search, this occurs when the probe face matches a face in the database and is one of the faces returned in the gallery of potential matches. In a one-to-one

rejection),[42] an inaccurate match (i.e., a false positive, or false alarm),[43] or an inaccurate non-match (i.e., a false negative, or miss).[44] It is the two types of errors—inaccurate matches and inaccurate non-matches—that have been of particular interest to policymakers.

- An inaccurate match, or *false positive,* result occurs when there is an erroneous association between images from two different people, which can occur when the digitized faces of two different people are highly similar.
- An inaccurate non-match, or *false negative,* result occurs when there is a failure to match images of the same person in two different photos. This could occur due to factors such as a change in the person's appearance or discrepancies in the quality of the images. Variations in pose, illumination, and expression may contribute to false negatives.[45]

Notably, there are both technical and human factors that contribute to the overall accuracy of facial recognition searches as performed by law enforcement officers.

Matching a probe to a gallery of images or a reference image depends on the threshold set for the similarity scores generated by the facial recognition algorithm. Similarity scores indicate the similarity between the probe and reference or gallery images.[46] For example, if using a zero-to- one scale, a

---

identity verification, this occurs when the probe face submitted matches a photo of the same individual in a database, and a match is confirmed.

[42] In a one-to-many identification search, this occurs when the probe face does not match a face in the database and a gallery of potential matches is not returned. In a one-to-one identity verification, this occurs when the probe face does not match any faces in a database, and a match is not confirmed.

[43] In a one-to-many identification search, this occurs when the probe face does not match a face in the database, but a gallery of potential matches is returned. In a one-to-one identity verification, this occurs when the probe face does not match a face in the database, but a match is confirmed.

[44] In a one-to-many identification search, this occurs when the probe face submitted matches a face in the database, but a gallery of potential matches is not returned. In a one-to-one identity verification, this occurs when the probe face matches a face in the database, but a match is not confirmed.

[45] See NIST, *Face Recognition Vendor Test (FRVT), Part 3: Demographic Effects (NIST IR8280)*, December 19, 2019, https://doi.org/10.6028/NIST.IR.8280.

[46] Similarity scores are sometimes referred to as *confidence scores*, but they do not represent a degree of certainty or confidence in the matches returned to the user or the accuracy of the system.

similarity score of one would indicate that the two images are most similar (not necessarily that the two face images belong to the same person) in that system. Further, similarity scores are system specific (i.e., a similarity score from a system developed by company A is not necessarily comparable to a similarity score from a system from company B).[47]

When trying to decide whether a probe image matches any images in a given database, setting a higher threshold will return fewer potential results and setting a lower threshold will return a greater number of potential results. Generally, the threshold is initially set by the algorithm developer. Depending on the system, the user can choose to keep or change this threshold. As with similarity scores, thresholds do not indicate accuracy of a system (i.e., adjusting a threshold to a higher value does not mean the results returned are more accurate); rather, the decision of where to set a threshold is based on how the system is being used and what the developer or user wants to optimize (e.g., reducing the chance of false positives or false negatives).[48] Notably, when considering where to set the threshold, there is also consideration of the inherent tradeoff in error rates—one can minimize the false positive rate or the false negative rate by resetting the threshold, but not both at the same time.[49]

There is also a range of possible accurate and inaccurate outcomes when the probe face and any gallery of potential matches are subject to facial review by a human. For instance, in one-to-many searches a reviewer can correctly match the probe face to the same individual's photo returned in the gallery of potential matches, or a reviewer can correctly reject the probe face as a match to faces in the gallery if the FRT software has returned a gallery that does not contain a match. In addition, a reviewer can incorrectly identify the probe face as a match to a face in a returned gallery (in either a gallery that correctly contains or incorrectly does not contain a match); alternatively, a reviewer could fail to identify the probe face as a match when a gallery contains the correct match. In one-to-one identity verifications, such as those used to confirm a traveler's identity, there may be follow up facial comparison by an

---

[47] For additional explanations of match scores and thresholds, see Partnership on AI, *Understanding Facial Recognition Systems*, February 19, 2020, pp. 6-7.

[48] Ibid.

[49] This tradeoff is demonstrated by plots that incorporate false negative and false positive identification rates with a threshold value; these plots are called detection error tradeoff (DET) characteristic or receiver operating characteristic (ROC) curves. For additional information see the 2019 NIST FRVTPart 3: Demographic Effects report, p. 23; and for further discussion, see Lucas D. Introna and Helen Nissenbaum, *Facial Recognition Technology: A Survey of Policy and Implementation Issues*, The Center for Catastrophe Preparedness and Response, New York University, 2009, pp. 14-15.

official in the instance of a no-match returned by the technology. In this case, the comparison can result in one of the same outcomes as one-to-many searches subject to follow-up review by a human.

Effects of Errors

False positives and negatives returned by FRT have come under scrutiny because of their potential implications. In one-to-many identification searches used by law enforcement, false positives could potentially contribute to errant investigative leads and false accusations. False negatives could potentially result in loss of evidence that could support a case. In one-to-one verifications used by border officials, false positives pose potential security risks because they may not flag a traveler using an assumed identity. False negatives could result in enhanced questioning, surveillance, or disrupted travel of individuals for whom it was not necessary.[50]

According to CBP internal analysis, the estimated false positive rate of TVS is .0103%.[51] (It did not report the false negative rate.) Further, in September 2020, the Government Accountability Office (GAO) reported that CBP "met or exceeded" its facial recognition accuracy requirements for its air exit system.[52]

A December 2019 NIST study of both one-to-many identification search algorithms and one-to- one verification algorithms found that FRT algorithms' accuracy rates can vary by demographic factors such as age, sex, and race.[53] For example, false positive rates tended to be higher for Asian and African American faces compared to those of Caucasians, which may be due to the data used to train the algorithm; an explanation that the NIST study did not explore. However, NIST noted that there is wide variation among algorithms, with some producing significantly fewer errors, and errors of different types,

---

[50] As a 2019 NIST study notes, "in a one-to-one access control, false negatives inconvenience legitimate users; false positives undermine a system owner's security goals. On the other hand, in a one-to-many deportee detection application, a false negative would present a security problem, and a false positive would flag legitimate visitors." Patrick Grother, Mei Ngan, and Kayee Hanaoka, Face Recognition Vendor Test (FRVT) Part 3: Demographic Effects, National Institute of Science and Technology, December 2019.

[51] Department of Homeland Security (DHS), Transportation Security Administration and U.S. Customs and Border Protection: Deployment of Biometric Technologies, Report to Congress, August 30, 2019, p. 30. CBP does not provide the methodology for calculating the false positive rate.

[52] U.S. Government Accountability Office (GAO), Facial Recognition: CBP and TSA are Taking Steps to Implement Programs, but CBP Should Address Privacy and System Performance Issues, GAO-20-568, September 2020, p. 50.

[53] Patrick Grother, Mei Ngan, and Kayee Hanaoka, *Face Recognition Vendor Test (FRVT) Part 3: Demographic Effects*, National Institute of Science and Technology, December 2019.

than others.[54] Policymakers may wish to exercise oversight over the specific FRT algorithms employed by federal law enforcement agencies, and the data on which those systems are trained, as they evaluate the accuracy and use of facial recognition. They may also debate whether or how to provide legislative direction aimed at maximizing the accuracy of FRT algorithms used by federal law enforcement entities. In attempting to maximize accuracy, developers and users of FRT must weigh the consequences of errors (false positives and false negatives) for different communities and decide which error measure is of higher priority to minimize, depending on how the threshold is set.

NIST researchers and collaborators have also studied the facial recognition accuracy of forensic examiners, superrecognizers, and face recognition algorithms.[55] They found that while the "best machine performed in the range of the best-performing humans, who were professional facial examiners … optimal face identification was achieved only when humans and machines collaborated."[56] Policymakers may consider this as they evaluate the accuracy of law enforcement use of FRT—such as the FBI's NGI-IPS, which requires manual review of the gallery of faces produced by submitting a probe face to the FRT algorithm.

Potential Restrictions on Law Enforcement Use of FRT

Recent policy debates surrounding law enforcement agencies' use of FRT have included discussions about potential prohibitions, restrictions, or moratoriums on the technology's use. In these discussions, policymakers may consider issues such as the following:

- *How is law enforcement conceptualized in this context?* As noted, law enforcement agencies with various missions—from those like

---

[54] Ibid. Also, testimony by Charles H. Romine, NIST Director, before U.S. Congress, House Committee on Homeland Security, *About Face: Examining the Department of Homeland Security's Use of Facial Recognition and Other Biometric Technologies: Part II*, 116th Cong., 2nd sess., February 6, 2020.

[55] P. Jonathon Phillips, Amy N. Yates, and Ying Hu et al., Proceedings of the National Academy of Sciences, *Face Recognition Accuracy of Facial Examiners, Superrecognizers, and Face Recognition Algorithms*, April 2018. Researchers note that superrecognizers are "untrained people with strong skills in face recognition."

[56] Testimony by Charles H. Romine, NIST, before U.S. Congress, House Committee on Oversight and Reform, *Facial Recognition Technology: Ensuring Transparency in Government Use*, 116th Cong., 1st sess., June 4, 2019.

the FBI's to investigate violations of federal criminal law to those like CBP's to support border enforcement—have employed FRT. Policymakers may consider whether proposals to specify whether or how law enforcement agencies may use FRT have also factored in which type of law enforcement activities might be affected.[57]

- *How might restrictions on the use of FRT affect emergencies or cases involving threats to national security?* Policymakers debating bounds on law enforcement agencies' use of FRT may consider whether restrictions should apply equally in all circumstances. For example, while many tools and technologies used by law enforcement agencies to aid investigations have not been specifically permitted or prohibited by law, Congress has legislated on and conducted oversight of certain technologies that could affect individual privacy. With electronic surveillance, for instance, investigators must generally obtain a warrant to conduct wiretaps;[58] however, exceptions exist for emergency situations that may involve death or serious injury, threaten national security, or involve conspiracies of organized crime.[59]
- *How might policymakers influence law enforcement use of FRT at the federal level as well as state and local levels?* Policymakers can legislate directly on federal law enforcement agencies' ability to utilize facial recognition and other biometric technologies, as well as specify under which circumstances federal law enforcement may use these tools. They can also direct federal departments and agencies to develop or rely on established guidelines surrounding the technologies, require them to use technology and FRT algorithms that meet specified standards, and conduct broad oversight of agencies' use of FRT.

Congress could also influence state, local, and tribal use of these technologies through the provision or withholding of grant funding. Programs such as the Edward Byrne Memorial Justice Assistance Grant (JAG)

---

[57] The Bureau of Justice Statistics' Census of Federal Law Enforcement Officers may help inform this discussion. The most recent survey provides data from 2016. Bureau of Justice Statistics, *Federal Law Enforcement Officers, 2016- Statistical Tables*, October 2019.

[58] 18 U.S.C. §2510, et seq. See also Department of Justice, *Justice Manual, Title 9, 9.7000: Electronic Surveillance.*

[59] 18 U.S.C. §2518.

program[60] and the Community Oriented Policing Services (COPS) program[61] have been used to incentivize activities of state and local law enforcement and may be leveraged to support or restrict agencies' use of FRT. For instance, Pinellas County, FL, law enforcement has used COPS funding to develop a facial recognition system.[62]

Another way the federal government can affect state, local, and tribal policies, without the provision or withholding of grant funding, is through the transfer of knowledge and expertise— via training, research and guiding documents, and model legislation. For instance, the Bureau of Justice Assistance published a guidance document for state, local, and tribal criminal intelligence and investigative entities to aid in developing policies around the use of FRT.[63]

Privacy and Security

In a September 2019 survey by the Pew Research Center, 56% of surveyed Americans indicated that they trust law enforcement agencies to use FRT responsibly, and 59% indicated it is acceptable for law enforcement agencies to use these technologies to assess security threats in public.[64] Further, the American public generally has more trust in law enforcement agencies using FRT responsibly than it does in technology companies or advertisers. This trust, however, has some notable demographic variances. Older Americans indicated they had more trust in law enforcement using FRT responsibly than did younger Americans. Further, White respondents (61%) reported more trust in law enforcement using the technology responsibly than did Hispanic respondents (56%), who in turn reported more trust than Black respondents

---

[60] For more information, see CRS In Focus IF10691, The Edward Byrne Memorial Justice Assistance Grant (JAG) Program.

[61] For more information, see CRS In Focus IF10922, Community Oriented Policing Services (COPS) Program.

[62] National Law Enforcement and Corrections Technology Center, Florida Facial Recognition System Unmasks Identity, Boosts Arrests, August 2010. Pinellas County, FL, law enforcement agencies had previously received, through the FY2001 appropriations, $3.5 million in funding for a demonstration grant "to demonstrate with the Florida Department of Motor Vehicles how facial recognition technology may be used by police." See H.Rept. 106-1005.

[63] Bureau of Justice Assistance, Face Recognition Policy Development Template, December 2017.

[64] Aaron Smith, Pew Research Center, More Than Half of U.S. Adults Trust Law Enforcement to Use Facial Recognition Responsibly, September 5, 2019.

(43%). Nonetheless, policymakers, advocates, and the public have raised questions about how these technologies might affect privacy as well as the security of facial recognition systems' data.[65] Questions for policymakers to consider include the following:

Is there public awareness and notification surrounding federal law enforcement use of FRT? Some have questioned whether or how individuals might know that their faces are included in databases searched by FRT for law enforcement purposes. For example, federal law enforcement agencies may rely on FRT to search against a number of databases to help identify suspects.[66] The FBI's FACE Services Unit can search probe photos against faces in NGI-IPS as well as other federal and state facial recognition systems authorized for FBI use. Some states allow FBI access to driver's license/identification photos, mugshot photos, and state department of corrections photos; some allow access to some portion or subset of those photos; and some prohibit access.[67] The FBI is required to provide information to the public on its facial recognition systems through Privacy Impact Assessments (PIAs) and System of Records Notices (SORNs).[68] Policymakers may question whether these are sufficient measures to notify the public about federal law enforcement agencies use of FRT to search against databases in which individuals' photos are held. In addition, they may conduct oversight

---

[65] The issues discussed in this section are focused on FRT as used by law enforcement to generate potential investigative leads and by border enforcement for identity verification. There are other potential uses of FRT in the criminal justice system, not discussed here, such as compelling an individual to use the facial recognition feature to unlock a mobile device such as an iPhone. Of note, "at least one court has upheld compelled use of a facial recognition unlock feature." See Joey L. Blanch and Stephanie S. Christensen, "Biometric Basics: Options to Gather Data from Digital Devices Locked by a Biometric "Key"," *Emerging Issues in Federal Prosecutions*, vol. 66, no. 1 (January 2018), p. 6.

[66] The same may be true for state and local law enforcement, but this varies by jurisdiction.

[67] GAO, *Face Recognition Technology: DOJ and FBI Have Taken Some Actions in Response to GAO Recommendations to Ensure Privacy and Accuracy, But Additional Work Remains*, GAO-2019-579T, June 4, 2019.

[68] Ibid. This requirement is not specific to the FBI. Federal agencies are subject to requirements under Section 208 of the E-Government Act of 2002 (P.L. 107-347) regarding the protection of personal information collected, maintained or disseminated using information technology. For more information, see Office of Management and Budget, *OMB Guidance for Implementing the Privacy Provisions of the E-Government Act of 2002*, M-03-22, September 26, 2003. In this guidance, a PIA is defined as "an analysis of how information is handled: (i) to ensure handling conforms to applicable legal, regulatory, and policy requirements regarding privacy, (ii) to determine the risks and effects of collecting, maintaining and disseminating information in identifiable form in an electronic information system, and (iii) to examine and evaluate protections and alternative processes for handling information to mitigate potential privacy risks." In addition, SORN is required to be published for any newly created or revised system of records.

over the timeliness with which federal law enforcement agencies publish and update relevant PIAs and SORNs.

There are also concerns about CBP's use of FRT. U.S. citizens are allowed to opt out of TVS biometric exit participation and can instead undergo manual review of travel documents. CBP notifies travelers of this alternative through its website, physical signs and verbal announcements at the ports of entry, and an FAQ sheet upon request.[69] However, a September 2020 GAO report found that "notices to inform the public of facial recognition contained limited privacy information and were not consistently available."[70] Policymakers may examine whether CBP provides U.S. citizens with adequate notice about TVS and explains its opt-out procedures clearly.[71]

How might the use of FRT affect police-community relations? In the national debate about police-community relations,[72] there have been concerns about whether race, gender, and age biases in some FRT algorithms could contribute to tensions between the police and the communities they serve. Further, in the midst of these discussions, some companies producing facial recognition software have decided to cease production or have enacted new barriers to law enforcement use of their technologies.[73] Some have been less concerned with potential errors produced by the technology and more apprehensive with how the technology may be *used* by law enforcement; specifically, the concern is whether law enforcement agencies' use of the technology can "strip individuals of their privacy and enable mass surveillance."[74]

Policymakers may continue to question whether federal law enforcement agencies have assessed—or have policies for ongoing assessments of—potential biases in the specific facial recognition technologies (and associated

---

[69] DHS, *Privacy Impact Assessment for the Traveler Verification Service,* DHS/CBP/PIA-056, November 14, 2018, p. 19.

[70] GAO, Facial Recognition: CBP and TSA are Taking Steps to Implement Programs, but CBP Should Address Privacy and System Performance Issues, GAO-20-568, September 2020, p. 39.

[71] Letter from 23 Members of Congress to Kevin McAleenan, former Acting Secretary of Homeland Security, June 13, 2019, https://wild.house.gov/sites/wild.house.gov/files/CBP%20Facial%20Recognition%20Ltr.%20final.%20.pdf.

[72] For more information, see CRS Report R43904, Public Trust and Law Enforcement—A Discussion for Policymakers.

[73] See, for example, Jay Greene, "Microsoft Won't Sell Police Its Facial-Recognition Technology, Following Similar Moves by Amazon and IBM," The Washington Post, June 11, 2020. See also Dustin Gardiner, "California Blocks Police From Using Facial Recognition in Body Cameras," San Francisco Chronicle, October 8, 2019.

[74] Osonde A. Osoba and Douglas Yeung, Bans on Facial Recognition Are Naive. Hold Law Enforcement Accountable for Its Abuse, RAND, June 17, 2020.

algorithms) that they use. This could include policies for ongoing assessments by NIST. Policymakers may also look into whether federal grants for state, local, and tribal law enforcement use of FRT include requirements that grantees are using facial recognition technologies that have been assessed for biases. In addition, they could continue to examine how federal law enforcement agencies, as well as state, local, and tribal recipients of federal grants, utilize the technology in their policing.

How do federal law enforcement agencies employing FRT retain and secure the data? The security of data held by federal agencies and their contractors is of ongoing interest to Congress.

For instance, in June 2019, CBP revealed that images of faces and license plates were compromised in a cyberattack on one of its subcontractors that provides automated license plate recognition technology to the agency.[75] This breach reportedly exposed confidential agreements, hardware schematics, and other records related to border security.[76] Breaches like this highlight the vulnerability of data, including face image data captured and held by governmental agencies.

In evaluating the security of federal law enforcement data systems, policymakers may pay particular attention to the security of facial recognition and other biometric data.

For example, the FBI's NGI-IPS contains mugshot photos against which probe photos are compared. The FBI notes that "after the facial recognition search is performed, the probe photo is not retained in the NGI-IPS. This ensures that the Criminal Identity Group of the NGI-IPS remains a repository of mugshots collected pursuant to arrest."[77] While NGI-IPS is often used by state, local, and tribal law enforcement agencies, the FACE Services Unit supports FBI investigations. The FACE Services Unit submits probe photos to NGI-IPS (which does not retain probe photos) as well as other federal, state, and local systems (which may have different policies on photo retention); when the FBI submits probe photos to entities outside the bureau, it is the other agency that is responsible for conducting the search. The FBI notes that "to accommodate certain states that have auditing and/or logging requirements that necessitate retention of probe photos and candidate galleries, the FBI

[75] For more information, see CRS Insight IN11143, Exposed Data Highlights Law Enforcement Use of Selected Technologies.

[76] Drew Harwell, "Surveillance Contractor That Violated Rules by Copying Traveler Images, License Plates Can Continue to Work with CBP," The Washington Post, October 10, 2019.

[77] FBI, Privacy Impact Assessment for the Next Generation Identification – Interstate Photo System, October 29, 2019, p. 2.

constructs [memoranda of understanding] in compliance with these requirements while also requiring state maintenance of only the minimum information necessary, for the shortest time period necessary, and notification to the FBI of any potential or actual breach of that information."[78]

CBP stores photographs of foreign nationals and U.S. citizens differently. All photographs are purged from the TVS cloud after 12 hours, regardless of citizenship status.[79] However, CBP stores photographs of foreign nationals for 14 days in the Automated Targeting System (ATS) Unified Passenger Module (UPAX).[80] These photographs are then transmitted to the Automated Biometric Identification System (IDENT), where they are retained for up to 75 years.[81] In contrast, photographs of U.S. citizens are immediately deleted after the matching process.[82] While CBP requires its commercial partners to follow these data retention requirements, a September 2020 GAO report found that CBP does not adequately audit its airline partners.[83]

Going Forward

There are currently no federal laws specifically governing law enforcement agencies' use of FRT,[84] and law enforcement agencies around the country may rely on a patchwork of technology platforms and algorithms for their facial recognition systems. As such, policymakers may question how federal law enforcement agencies assess and ensure the accuracy and security of their FRT systems as well as the policies governing their use. They may also examine the oversight of federal grants used to support or restrict the use of FRT by state, local, and tribal law enforcement. Policymakers may further consider how FRT is used more broadly in various phases of the criminal justice system, from generating leads and helping establish probable cause for an

---

[78] FBI, Privacy Impact Assessment for the Facial Analysis, Comparison, and Evaluation (FACE) Phase II System, July 9, 2018, p. 8.

[79] Ibid., p. 9.

[80] DHS, Privacy Impact Assessment for the Traveler Verification Service, DHS/CBP/PIA-056, November 14, 2018, pp. 9, 21.

[81] Ibid., pp. 8, 21.

[82] Ibid., p. 10.

[83] GAO, Facial Recognition: CBP and TSA are Taking Steps to Implement Programs, but CBP Should Address Privacy and System Performance Issues, GAO-20-568, September 2020, p. 46.

[84] For a discussion of relevant constitutional considerations surrounding law enforcement use of FRT, see CRS Report R46541, Facial Recognition Technology and Law Enforcement: Select Constitutional Considerations.

arrest or indictment, to serving as evidence in courtrooms and confirming prisoners' identities before release; this may help inform oversight and legislative efforts to enhance or bound aspects of how law enforcement uses the technology. In addition, policymakers may question whether or how recommendations from FISWG are adopted by federal law enforcement agencies; their state, local, and tribal partners; and law enforcement grant recipients.

Appendix. NIST Efforts on Facial Recognition Technology

NIST work on FRT includes research, testing and evaluation, technical guidance and support, and standards.[85]

Research

NIST work in biometrics dates back to the 1960s. The agency's efforts span a wide range of activities to help improve the ability to establish or verify the identity of humans based upon one or more physical (e.g., face, fingerprint, iris images) or behavioral (e.g., signature analysis) characteristics.

The Information Technology Laboratory (ITL), one of six NIST research laboratories, is a measurement and testing facility that develops and deploys standards, tests, and metrics to make information systems more secure, usable, interoperable, and reliable. Among its functions, ITL conducts research on issues related to biometric measurement and testing and facilitates standards development, including those related to FRT. According to NIST, ITL has measured the core algorithmic capability of biometric recognition technologies and reported on the accuracy, throughput, reliability, and sensitivity of biometric algorithms with respect to data characteristics and subject characteristics.

NIST states that its biometric evaluations advance measurement science by providing a scientific basis for what to measure and how to measure it. These evaluations also help facilitate development of consensus-based standards by providing quantitative data for development of scientifically

[85] Much of the information in this appendix is drawn from testimony given on February 6, 2020, by Charles H. Romine, Director of the National Institute of Standards and Technology's Information Technology Laboratory before the House Committee on Homeland Security.

sound, fit-for-purpose standards. In addition, these evaluations help federal agencies determine how best to deploy FRT.

NIST's FRT research includes a wide span of activities as illustrated by the following examples:

In 2018, the National Academies published research conducted by NIST and three universities testing facial forensic examiners ability to match identities across different photographs.[86] The intent of the study was to find better ways to increase the accuracy of forensic facial comparisons. The study concluded that

> Examiners and other human face "specialists," including forensically trained facial reviewers and untrained super-recognizers, were more accurate than the control groups on a challenging test of face identification. It also presented data comparing state-of-the-art facial recognition algorithms with the best human face identifiers. The best machine performed in the range of the best-performing humans, who were professional facial examiners. However, optimal face identification was achieved only when humans and machines collaborated.[87]

In addition, NIST conducted the Face in Video Evaluation (FIVE) program to assess the capability of facial recognition algorithms to identify individuals appearing in video sequences. NIST documented the outcomes of FIVE in its report, Face In Video Evaluation (FIVE) Face Recognition of Non-Cooperative Subjects (NIST IR8173), which discusses the accuracy and speed of FRT algorithms applied to the identification of individuals appearing in video sequences drawn from six video datasets.[88] NIST completed this program in 2017. The report found that

> High accuracy recognition of passively-imaged subjects is only achievable with: a) a small minority of the algorithms tested [under this program]; b) a dedicated and deliberate design effort that must embed

[86] Jonathon Phillips, Amy N. Yates, and Ying Hu, "Face recognition accuracy of forensic examiners, superrecognizers, and face recognition algorithms," *Proceedings of the National Academy of Sciences*, vol. 115, no. 24 (June 12, 2018), pp. 6171-6176, https:// doi.org/10. 1073/pnas.1721355115.

[87] Testimony of Charles H. Romine, Director, NIST Information Technology Laboratory, before U.S. Congress, House Committee on Homeland Security, Facial Recognition Technology (FRT), 116th Cong., 2nd sess., February 6, 2020, https://www.nist.gov/speech-testimony/ facial-recognition-technology-frt-0 (hereinafter, Romine FRT testimony).

[88] NIST, Face In Video Evaluation (FIVE) Face Recognition of Non-Cooperative Subjects (NIST IR8173), March 2017, https://doi.org/10.6028/NIST.IR.8173.

> optical, architectural, human factors, operations-research, and face recognition expertise; c) galleries limited to small numbers of actively curated images; and d) field tests with empirical quantitative calibration and optimization.[89]

Further, the report states that with "better cameras, better design, and the latest algorithm developments, recognition accuracy can advance even further," but notes that "even with perfect design, some proportion of a non-cooperative population will not be recognized" due to failure to acquire cases where subjects never look toward the camera or because their faces were occluded.[90]

The report concluded, in part, that

> Deployment should proceed only after quantitative assessment of objectives, alternatives, ease of evasion or circumvention, enrolled population sizes, search volumes, the proportion of searches expected to have an enrolled mate, accuracy requirements, consequences and procedures for resolution of errors, and speed and hardware cost constraints. In particular, deployers must weight their tolerance for misses and their risk appetite. In addition, when non-cooperative face recognition is used to identify individuals nominated to a watchlist, human reviewers must be employed to adjudicate whether candidate matches are true or false positives … [and that] overall error rates of the hybrid machine-human system must be understood and planned for.[91]

NIST has also conducted a number of FRT-related "Grand Challenge" competitions—including the Face Recognition Grand Challenge (2004-2006) and the Multiple Biometric Grand Challenge (2008-2010) programs—to encourage the FRT community to break new ground in solving biometric research problems.

Testing and Evaluation

Since 2000, NIST has operated a Face Recognition Vendor Testing (FRVT) program to assess the capabilities of facial recognition algorithms for one-to-

---

[89] Ibid.
[90] Ibid.
[91] Ibid.

many identification and one-to-one verification. The voluntary program is open to any organization worldwide, and participants may submit their algorithms on a continuous basis for evaluation. Users include corporate research and development laboratories and universities. Submitted algorithms include commercially available products and prototypes that are not necessarily available as final products ready for integration in FRT systems. NIST posts performance results for evaluated algorithms on its FRVT website along with the name of the organization that developed the algorithm.[92]

According to NIST, the FRVT program does not train face recognition algorithms.[93] NIST does not provide training data to the software under test, and the software is prohibited from adapting to any data that is passed to the algorithms during a test.[94]

With respect to its 2019 FRVT activities, NIST reported that

> The 2019 FRVT quantified the accuracy of face recognition algorithms for demographic groups defined by sex, age, and race or country of birth, for both one-to-one verification algorithms and one-to-many identification search algorithms. NIST conducted tests to quantify demographic differences for 189 face recognition algorithms from 99 developers, using four collections of photographs with 18.27 million images of 8.49 million people. These images came from operational databases provided by the State Department, the Department of Homeland Security and the FBI. Previous FRVT reports documented the accuracy of these algorithms and showed a wide range in accuracy across algorithms. The more accurate algorithms produce fewer errors and can therefore be anticipated to have smaller demographic differentials.
>
> NIST Interagency Report 8280,[95] released on December 19, 2019, quantifies the effect of age, race, and sex on face recognition

[92] NIST, "FRVT 1:1 Leaderboard," https://www.nist.gov/programs-projects/face-recognition-vendor-test-frvt. The FRVT 1:1 Leaderboard shows the top performing 1:1 algorithms measured on false non-match rate across several different datasets.

[93] According to NIST, "The process of training a face recognition algorithm (or any machine learning algorithm) involves providing a machine learning algorithm with training data to learn from. The training data shall contain the correct answer, which is known as ground-truth label, or a target. The learning algorithm finds patterns in the training data that map the input data attributes to the target and builds a machine-learning model that captures these patterns. This model can then be used to get predictions on new data for which the target is unknown." See Romine FRT testimony.

[94] Romine FRT testimony.

[95] NIST, *Face Recognition Vendor Test (FRVT), Part 3: Demographic Effects (NIST IR8280)*, December 19, 2019, https://doi.org/10.6028/NIST.IR.8280. This is the third in a series of reports on the 2019 FRVT activities that extends the evaluations from parts 1 and 2—which

> performance. It found empirical evidence for the existence of demographic differentials in face recognition algorithms that NIST evaluated. The report distinguishes between false positive and false negative errors, and notes that the impacts of errors are application dependent.[96]

In interpreting the results of the 2019 FRVT report on demographic effects, one should note that the study used high-quality, standards-compliant images, not image data from the Internet nor from video surveillance. Thus, demographic differentials from images in such everyday scenarios were not evaluated. Some stakeholders have criticized such limitations in analyzing FR algorithms and called for more testing "in the field under real-life conditions."[97]

With respect to its 2018 FRVT, NIST reported that

> The 2018 FRVT tested 127 facial recognition algorithms from the research laboratories of 39 commercial developers and one university, using 26 million mugshot images of 12 million individuals provided by the FBI. The 2018 FRVT measured the accuracy and speed of one-to-many facial recognition identification algorithms. The evaluation also contrasted mugshot accuracy with that from lower quality images. The findings, reported in *NIST Interagency Report 8238*,[98] showed that massive gains in accuracy have been achieved since the FRVT in 2013, which far exceed improvements made in the prior period (2010- 2013). The accuracy gains observed in the 2018 FVRT study stem from the integration, or complete replacement, of older facial recognition techniques with those based on deep convolutional neural networks. While the industry gains are broad, there remains a wide range of capabilities, with some developers providing much more accurate algorithms than others do. Using FBI mugshots, the most accurate algorithms fail only in about one quarter of one percent of searches, and

---

covered the performance of one-to-one and one-to-many face recognition algorithms, respectively—to document accuracy variations across demographic groups.

96 Romine FRT testimony.

97 Darrell M. West, *10 Actions That Will Protect People From Facial Recognition Software,* Brookings report, October 31, 2019, https://www.brookings.edu/research/10-actions-that-will-protect-people-from-facial-recognition-software/.

98 NIST, Ongoing Face Recognition Vendor Test (FRVT) Part 2: Identification (NIST IR8238), November 2018, https://doi.org/10.6028/NIST.IR.8238. This report was subsequently updated and extended by NIST as Face Recognition Vendor Test (FRVT) Part 2: Identification, NIST Interagency Report 8271, September 2019, https://doi.org/10.6028/NIST.IR.8271.

these failures are associated with images of injured persons and those with long time lapse since the first photograph. The success of mugshot searches stems from the new generation of facial recognition algorithms, and from the adoption of portrait photography standards first developed at NIST in the late 1990s.[99]

Technical Guidance and Scientific Support

NIST provides technical guidance and scientific support to various U.S. government and law enforcement agencies for the use of FRT. For example, NIST's research supported DHS's transition from a 2-fingerprint to a 10-fingerprint collection standard for visa application and entry into the United States to prevent terrorism and identity fraud as well as to prevent criminals and immigration violators from crossing U.S. borders. In addition, NIST is working with CBP to analyze performance effects of image quality and traveler demographics, and to provide recommendations regarding match algorithms, optimal thresholds, and match gallery creation for TVS. NIST's work also supports the FBI and the Office of the Director of National Intelligence's Intelligence Advanced Research Projects Activity (IARPA).[100]

Standards

The United States has a voluntary, consensus-based standards development system.[101] Under the National Technology Transfer and Advancement Act of

[99] Romine FRT testimony.

[100] Romine FRT testimony.

[101] According to the American National Standards Institute (ANSI):

The U.S. standardization system reflects a market-driven and highly diversified society. It is a decentralized system that is naturally partitioned into industrial sectors and supported by independent, private sector standards developing organizations (SDOs). It is a demand-driven system in which standards are developed in response to specific concerns and needs expressed by industry, government, and consumers. And it is a voluntary system in which both standards development and implementation are driven by stakeholder needs.... Voluntary standards serve as the cornerstone of the distinctive U.S. infrastructure. These documents arise from a formal, coordinated, consensus-based and open process. Their development depends upon data gathering, a vigorous discussion of all viewpoints, and agreement among a diverse range of stakeholders....

Voluntary refers only to the manner in which the standard was developed; it does not necessarily refer to whether compliance to a consensus standard is optional or whether a government entity or market sector has endorsed the document for mandatory use. Most other countries

1995 (P.L. 104-113)[102] and OMB Circular A-119,[103] NIST is charged with promoting coordination between the public and private sectors in the development of standards and in conformity assessment activities, encouraging and coordinating federal agency use of voluntary consensus standards in lieu of government-unique standards, and coordinating federal agency participation in the development of relevant standards.

NIST leads national and international consensus standards activities in biometrics, such as FRT, to ensure that they are interoperable, reliable, secure, and usable.

The following examples of NIST consensus standards development activities illustrate NIST's role in this arena:

Data Format for the Interchange of Fingerprint, Facial & Other Biometric Information (ANSI/NIST-ITL 1-2011 Update: 2015), published by the American National Standards Institute and NIST, is a biometric standard used in 160 countries to facilitate the exchange of biometric data across jurisdictional lines and between dissimilar systems. This standard allows accurate and interoperable exchange of biometrics information by law enforcement agencies globally, assisting in the identification of criminals and terrorists. The standard continues to evolve to support government applications, including law enforcement, homeland security, and other identity management applications. According to NIST, the standard is used for nearly all law enforcement biometric collections worldwide.[104]

---

adhere to a "top- down" approach to standardization where the government or groups closely coupled to government either serve as the standards setter or mandate what standards will be developed.

(American National Standards Institute, "Overview of the U.S. Standardization System," https://www.standardsportal.org/usa_en/standards_system.aspx.) ANSI was founded in 1918 by five engineering societies and three federal agencies (the Departments of War, Navy, and Commerce). ANSI is the sole U.S. representative and dues-paying member of the two major non-treaty international standards organizations, the International Organization for Standardization (ISO) and the International Electrotechnical Commission (IEC).

102 Codified at 15 U.S.C. Section 272(b), which directs the Secretary of Commerce, through the NIST director, "to facilitate standards-related information sharing and cooperation between Federal agencies and to coordinate the use by Federal agencies of private sector standards, emphasizing where possible the use of standards developed by private, consensus organizations" and "to coordinate technical standards activities and conformity assessment activities of Federal, State, and local governments with private sector technical standards activities and conformity assessment activities, with the goal of eliminating unnecessary duplication and complexity in the development and promulgation of conformity assessment requirements and measures."

103 Executive Office of the President, Office of Management and Budget, Circulars, https://www.whitehouse.gov/omb/information-for-agencies/circulars/.

104 Romine FRT testimony.

NIST has also led and provided technical expertise for the development of international biometric standards in ISO/IEC Joint Technical Committee 1, Subcommittee 37 (JTC1/SC37) –Biometrics.[105] The standards developed by the subcommittee, which was established in 2002, are broadly used nationally and internationally. Since 2006, the subcommittee has published standards on biometric performance testing and reporting (including guidance on principles and framework, testing methodologies, modality-specific testing, interoperability performance testing, and access control scenarios), drawing upon NIST technical contributions.

[105] ISO is the International Organization for Standards. IEC is the International Electrotechnical Commission.

Chapter 2

Facial Recognition Technology: Current and Planned Uses by Federal Agencies*

United States Government Accountability Office

Why GAO Did This Study

Facial recognition—a type of biometric technology—mimics how people identify or verify others by examining their faces. Recent advancements have increased the accuracy of automated FRT resulting in increased use across a range of applications. As the use of FRT continues to expand, it has become increasingly important to understand its use across the federal government in a comprehensive way.

GAO was asked to review the extent of FRT use across the federal government. This chapter identifies and describes (1) how agencies used FRT in fiscal year 2020, including any related research and development and interactions with non-federal entities, and (2) how agencies plan to expand their use of FRT through fiscal year 2023.

GAO surveyed the 24 agencies of the Chief Financial Officers Act of 1990, as amended, regarding their use of facial recognition technology. GAO also interviewed agency officials and reviewed documents, such as system descriptions, and information provided by agencies that reported using the technology.

---

* This is an edited, reformatted and augmented version of United States Government Accountability Office, Publication No. GAO-21-526, dated August 2021.

In: Facial Recognition Technology
Editor: Mari F. Burke
ISBN: 979-8-88697-124-8

What GAO Found

In response to GAO's survey about facial recognition technology (FRT) activities in fiscal year 2020, 18 of the 24 surveyed agencies reported using an FRT system, for one or more purposes, including:

- *Digital access or cybersecurity.* Sixteen agencies reported using FRT for digital access or cybersecurity purposes. Of these, 14 agencies authorized personnel to use FRT to unlock their agency-issued smartphones—the most common purpose of FRT reported. Two agencies also reported testing FRT to verify identities of persons accessing government websites.
- *Domestic law enforcement.* Six agencies reported using FRT to generate leads in criminal investigations, such as identifying a person of interest, by comparing their image against mugshots. In some cases, agencies identify crime victims, such as exploited children, by using commercial systems that compare against publicly available images, such as from social media.
- *Physical security.* Five agencies reported using FRT to monitor or surveil locations to determine if an individual is present, such as someone on a watchlist, or to control access to a building or facility. For example, an agency used it to monitor live video for persons on watchlists and to alert security personnel to these persons without needing to memorize them.

Source: GAO analysis of survey results and GoldenSikora/metamorworks/Cipta/ stock.adobe.com. | GAO-21-526.

Examples of Facial Recognition Technology Uses by Federal Agencies.

Ten agencies reported FRT-related research and development. For example, agencies reported researching FRT's ability to identify individuals wearing masks during the COVID-19 pandemic and to detect image manipulation.

Furthermore, ten agencies reported plans to expand their use of FRT through fiscal year 2023. For example, an agency plans to pilot the use of FRT to automate the identity verification process at airports for travelers.

Abbreviations

CBP	U.S. Customs and Border Protection
CFO Act	Chief Financial Officers Act of 1990
COVID-19	Coronavirus Disease 2019
DHS	Department of Homeland Security
DOD	Department of Defense
DOE	Department of Energy
DOJ	Department of Justice
DOT	Department of Transportation
EPA	Environmental Protection Agency
FACE	Facial Analysis, Comparison, and Evaluation
FBI	Federal Bureau of Investigation
FRT	facial recognition technology
GSA	General Services Administration
HHS	Department of Health and Human Services
HSIN	Homeland Security Information Network
IDENT	Automated Biometric Identification System
NASA	National Aeronautics and Space Administration
NCRFRILS	National Capital Region Facial Recognition Investigative Leads System
NSF	National Science Foundation
OPM	Office of Personnel Management
PIN	personal identification number
RAPIDS	Real-time Automated Personnel Identification System
SSA	Social Security Administration
TSA	Transportation Security Administration
USAID	U.S. Agency for International Development
USDA	U.S. Department of Agriculture
VA	Department of Veterans Affairs

Of all the biometric technologies—those used to identify people based on their biological and behavioral characteristics—facial recognition most closely mimics how people identify others: by examining their faces. What is an effortless skill in humans has proven difficult to replicate in machines, but computer and technology advancements over the past few decades have increased the overall accuracy of automated facial recognition. As a result, the use of facial recognition technology (FRT) has become increasingly common across business and government sectors. For example, it is used as a tool for identifying or verifying customers, and to verify an employee's identity when logging into a computer. Law enforcement can also use it to search databases, such as driver's license photos and mugshots, for possible leads about an unknown individual's identity as part of a criminal investigation.

As the use of FRT continues to expand, Members of Congress, academics, and advocacy organizations have highlighted the importance of developing a comprehensive understanding of how it is used by federal agencies. This chapter is the latest in a series of recent reports we have issued on FRT. In June 2021, we reported on federal law enforcement's use of FRT, including the extent of its use and how the agencies monitor such use.[1] In September 2020, we reported on the U.S. Customs and Border Protection's (CBP) and Transportation Security Administration's (TSA) use of FRT at U.S. ports of entry and made recommendations to CBP to improve its privacy practices and system performance.[2] In 2016, we reported on the Federal Bureau of Investigation's (FBI) use of FRT and made recommendations to improve the Bureau's understanding of the accuracy of and privacy protection processes for its FRT capabilities.[3] We have also recently reported on the ways FRT can be used in commercial settings, including to provide secure access to online

---

[1] GAO, *Facial Recognition Technology: Federal Law Enforcement Agencies Should Better Assess Privacy and Other Risks*, GAO-21-518 (Washington, D.C., June 3, 2021). In this report, we made 26 recommendations related to using nonfederal systems with facial recognition technology to eight agencies, including the Departments of Health and Human Services, Homeland Security, the Interior, Justice, State, and the Treasury.

[2] GAO, *Facial Recognition: CBP and TSA are Taking Steps to Implement Programs, but CBP Should Address Privacy and System Performance Issues*, GAO-20-568 (Washington, D.C.: Sept. 2, 2020). In this report, we made five recommendations to CBP related to its use of facial recognition technology, and DHS concurred with the recommendations. In March and April 2021, CBP provided a status update on progress towards each of these recommendations. Based on the documentation provided by CBP, GAO closed two recommendations as implemented.

[3] GAO, *Face Recognition Technology: FBI Should Better Ensure Privacy and Accuracy*, GAO-16-267 (Washington, D.C.: May 16, 2016). In this report, we made six recommendations related to accuracy and privacy. The FBI has addressed all six recommendations.

customer accounts and information on customer flows during peak times, among others.[4]

You asked us to review the extent of FRT use across the federal government. This chapter identifies and describes (1) how agencies used FRT in fiscal year 2020, including any FRT-related research and development activities and interactions with nonfederal entities, and (2) how agencies plan to expand their use of FRT through fiscal year 2023.

To address these objectives, we administered a survey to the 24 Chief Financial Officers (CFO) Act agencies[5] to collect information related to: (1) FRT systems used (owned or accessed, including those tested) by agencies during fiscal year 2020; (2) FRT systems agencies planned to use through fiscal year 2023; (3) agencies' FRT-related research and development activities; (4) transactions (financial or otherwise) that agencies entered into for nonfederal entities' use of FRT; and (5) the extent to which agencies regulated nonfederal entities' use of FRT.[6] The questionnaire also asked detailed questions about the individual FRT systems that agencies reported, which included the purpose(s), a brief description of its use, and obligations related to its use. We asked agencies to include the activities of all their components, bureaus, and offices in their responses. We emailed questionnaires to the agencies in October 2020, and closed the survey in January 2021 after receiving responses from all 24 agencies. We reviewed the responses we collected and took quality control steps by performing checks for completeness, logical errors, and inconsistencies. We followed up with agencies in writing or through interviews, as appropriate. See appendix I for additional information on our scope and methodology.

---

[4] GAO, *Facial Recognition Technology: Privacy and Accuracy Issues Related to Commercial Uses*, GAO-20-522 (Washington, D.C.: July 13, 2020) and GAO, *Facial Recognition Technology: Commercial Uses, Privacy Issues, and Applicable Federal Law*, GAO-15-621 (Washington, D.C.: July 30, 2015).

[5] The 24 agencies are those identified in the Chief Financial Officers Act of 1990, as amended (31 U.S.C. § 901(b)). They are the U.S. Departments of Agriculture, Commerce, Defense, Education, Energy, Health and Human Services, Homeland Security, Housing and Urban Development, the Interior, Justice, Labor, State, Transportation, the Treasury, Veterans Affairs, Environmental Protection Agency, National Aeronautics and Space Administration, U.S. Agency for International Development, General Services Administration, National Science Foundation, Nuclear Regulatory Commission, Office of Personnel Management, Small Business Administration, and Social Security Administration.

[6] Regulated refers to using regulatory authority over a nonfederal entity to regulate that entity's use of its own FRT. For the purposes of our questionnaire, we defined "regulated" as regulatory functions in which the agency engaged, including, but not limited to, investigatory and inspections activities, taking enforcement actions, prescribing requirements or guidance, conducting oversight, and maintaining performance standards.

We conducted this performance audit from April 2020 through August 2021 in accordance with generally accepted government auditing standards. Those standards require that we plan and perform the audit to obtain sufficient, appropriate evidence to provide a reasonable basis for our findings and conclusions based on our audit objectives. We believe that the evidence obtained provides a reasonable basis for our findings and conclusions on our audit objectives.

Background

How Facial Recognition Technology Works

Facial recognition can verify or identify individuals by their faces. It is one of several biometric technologies that identify individuals by measuring and analyzing physical and behavioral characteristics.[7] As seen in Figure 1, facial recognition technology uses a photo or still from a video feed of a person—often called a probe or live photo—and converts it into a template, or a mathematical representation of the photo. A matching algorithm can then compare the template to one from another photo and calculate their similarity.

Facial recognition searches or comparisons generally fall into two categories: verification and identification. Verification (or one-to-one searches) compares a stored photo of an individual to another photo purportedly of the same individual to determine whether they are the same person. For example, this type of comparison can help verify the identity of an individual attempting to unlock a smartphone. Identification (or one-to-many searches) compares a photo from a single individual against a gallery of stored photos from a number of individuals to determine if there is a potential match. For example, this type of comparison can be used to identify investigative leads for an unknown individual in a crime scene photo.

Two technologies, facial detection and facial analysis, are related to, but distinct from, facial recognition. Whereas facial recognition matches a face to a specific identity:

- *Facial detection* determines if a photo or video contains a face in the image. It is commonly used to count the number of people that move

[7] Other biometric technologies can identify individuals by measuring and analyzing physical and behavioral characteristics, which include fingerprints, eye irises, voice, and gait.

through a particular area without determining their identities, such as counting people in stores or amusement park lines.

- *Facial analysis*, sometimes referred to as facial classification or characterization, uses a facial image to estimate or classify personal characteristics such as age, race, or sex, or tracks facial features or movement to recognize expressions or gaze, among other analyses. For example, facial analysis can be part of an eye tracking system, which can allow researchers to analyze how well pilots use their eyes or gaze to scan their cockpit instruments.

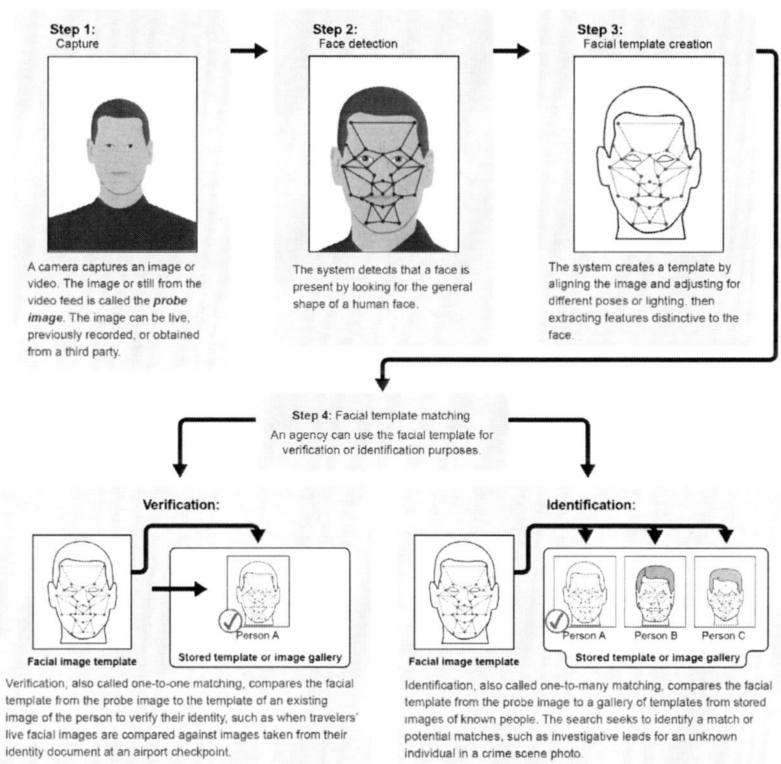

Source: GAO analysis. | GAO-21-526.

Figure 1. Process used in Facial Recognition Technology.

For the purposes of this chapter, we use the term "facial recognition technology" to include facial recognition, facial detection, or facial analysis technologies.

Federal Use of Facial Recognition Technology Systems

A facial recognition technology system may include components or modules of systems, software applications, or devices with automated facial recognition capabilities, such as a face recognition algorithm, hardware, or software. Federal agencies can own their FRT systems or access the FRT systems of other government entities, including federal, state, local, tribal, and territorial governments, and commercial facial recognition service providers.[8] Agencies can have direct access to an FRT system, such as by logging into the system, or indirect access, such as by requesting a state government (i.e., a third party) run a facial recognition search on behalf of the federal agency.

FRT systems are used for a variety of purposes across the federal government. These purposes and examples of how FRT can be used are grouped into seven different categories, as follows:[9]

- *Digital access or cybersecurity.* This purpose includes FRT that can be used to control access to a personal computer, smartphone, or mobile application.
- *Domestic law enforcement.* This purpose includes FRT that can be used to identify a lead or person of interest in an investigation, or to locate or identify a missing person or crime victim.
- *Physical security.* This purpose includes FRT that can be used to control physical access, such as to facilities or buildings, or to surveil or monitor a location or facility, including notification that an individual is present in real-time.
- *Border and transportation security.* This purpose includes FRT that can be used to confirm the identities of domestic travelers at airports, travelers applying to enter the United States or crossing U.S. borders, or non-citizens in immigration proceedings.
- *National security and defense.* This purpose includes FRT that can be used to research derogatory information on a known or suspected

---

[8] We use the term "federally owned" when an agency developed or acquired the FRT and performs its own searches. We use the term "commercially owned" when an agency contracts for an FRT service, such as a search performed against the commercial entity's database of images or performed by the commercial entity itself and the results are provided back to the agency.

[9] For the purposes of this report, we determined these categories to describe the different ways agencies could use FRT.

terrorist, or confirm the identity of a foreign national for national security reasons.[10]

- *Medical assessment.* This purpose includes FRT that can be used to confirm a patient's identity in a medical setting, such as when dispensing controlled substance prescriptions, or to assist with contact tracing (e.g., related to Coronavirus Disease 2019 or COVID-19).[11]
- *Other.* These FRT purposes include analysis of the face itself, such as analysis of attention or alertness based on eye tracking, or inferring characteristics of a person, including age or sex. It also includes any other agency use that does not fit into the categories above.

Illustrative examples of these purposes are shown in Figure 2.

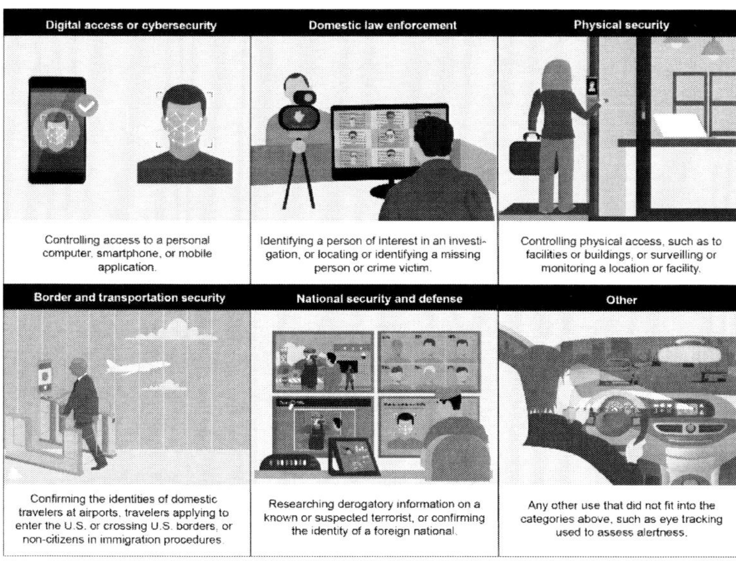

Source: GAO analysis of survey results and GoldenSikora/metamorworks/Cipta/stock.adobe.com. | GAO-21-526.

Figure 2. Examples of Federal Agencies' Use of Facial Recognition Technology (FRT).

---

[10] For example, to confirm the identity of a foreign national seeking access to a military installation.

[11] For information on how contact tracing can occur, see GAO, *Science & Tech Spotlight: Contact Tracing Apps*, GAO-20-666SP (Washington, D.C., Jul. 28, 2020).

Agencies Most Often Reported Using FRT for Digital Access and Domestic Law Enforcement

Most of the federal agencies we surveyed—19 of 24—reported one or more FRT-related activities in fiscal year 2020, with digital access and domestic law enforcement as the most common. For the purposes of our survey, we identified four types of FRT activities an agency can engage in: (1) using an FRT system, which includes owning, accessing, or testing the system; (2) conducting or supporting FRT-related research and development; (3) entering into transactions with nonfederal entities, such as awarding grants to enable entities to obtain FRT systems for their own uses; and (4) regulating the use of FRT by nonfederal entities. Table 1 shows the four types of FRT activities reported in fiscal year 2020 by the 24 federal agencies we surveyed. Five agencies reported they did not conduct any of these FRT activities in fiscal year 2020: the Department of Education, Department of Housing and Urban Development, Department of Labor, the Nuclear Regulatory Commission, and the Small Business Administration.

Eighteen Agencies Reported Using FRT for a Variety of Purposes

Eighteen of the 24 agencies we surveyed responded that they used facial recognition technology in fiscal year 2020, as shown in the first column of Table 1 below. These agencies reported using FRT for one or more purposes, with digital access and domestic law enforcement as the most common (see Table 2). Agencies did not report using FRT for medical assessment purposes in fiscal year 2020.

Examples of how agencies used FRT in each purpose category are described below.

Table 1. Facial recognition technology (FRT) activities reported by federal agencies for fiscal year 2020

Federal Agency	Used FRT systems[a]	Conducted FRT- related research and development[b]	Entered into transactions with nonfederal entities for FRT[c]	Regulated nonfederal entities' use of FRT[d]
Department of Agriculture	•	⊗	⊗	⊗
Department of Commerce	•	•	⊗	⊗
Department of Defense	•	•	⊗	⊗
Department of Education	⊗	⊗	⊗	⊗
Department of Energy	•	⊗	⊗	⊗
Department of Health and Human Services	•	•	⊗	⊗
Department of Homeland Security	•	•	•	•[e]
Department of Housing and Urban Development	⊗	⊗	⊗	⊗
Department of the Interior	•	⊗	⊗	⊗
Department of Justice	•	•	•	⊗
Department of Labor	⊗	⊗	⊗	⊗
Department of State	•	•	•	⊗
Department of Transportation	⊗	•	⊗	⊗
Department of the Treasury	•	⊗	⊗	⊗
Department of Veterans Affairs	•	•	•	⊗
U.S. Agency for International Development	•	⊗	⊗	⊗
Environmental Protection Agency	•	⊗	⊗	⊗
General Services Administration	•	⊗	⊗	⊗
National Aeronautics and Space Administration	•	•	⊗	⊗
National Science Foundation	•	•	⊗	⊗
Nuclear Regulatory Commission	⊗	⊗	⊗	⊗
Office of Personnel Management	•	⊗	⊗	⊗
Small Business Administration	⊗	⊗	⊗	⊗
Social Security Administration	•	⊗	⊗	⊗

• Yes; ⊗ No

Source: GAO analysis of survey results. | GAO-21-526.

[a]The agency owned or accessed FRT in fiscal year 2020.

[b]Research and development also includes agencies that obligated funds for FRT-related research conducted by other entities.

[c]Refers to agencies that entered into transactions, such as awarding grants to nonfederal entities to purchase FRT equipment.

[d]Refers to using regulatory authority over a nonfederal entity to regulate that entity's use of its own FRT. For the purposes of our questionnaire, we defined "regulated" as regulatory functions in which the agency engaged, including, but not limited to, investigatory and inspections activities, taking enforcement actions, prescribing requirements or guidance, conducting oversight, and maintaining performance standards.

[e]The Transportation Security Administration within the Department of Homeland Security used its authority to regulate entities under its jurisdiction.

Table 2. Reported purposes of facial recognition technology systems used by federal agencies in fiscal year 2020

	Purpose					
Federal Agency	Digital access	Domestic law enforcement	Physical security	Border and transportation security	National security and defense	Other
Department of Agriculture	•	⊗	⊗	⊗	⊗	⊗
Department of Commerce	•	⊗	•	⊗	⊗	⊗
Department of Defense	⊗	•	•	⊗	•	•
Department of Energy	•	⊗	•	⊗	⊗	⊗
Department of Health and Human Services	•	•	•	⊗	⊗	⊗
Department of Homeland Security	•	•	⊗	•	•	⊗
Department of the Interior	•	•	⊗	⊗	⊗	⊗
Department of Justice	•	•	•	⊗	•	•
Department of State	⊗	⊗	⊗	•	•	⊗
Department of the Treasury	•	•	⊗	⊗	⊗	⊗
Department of Veterans Affairs	•	⊗	⊗	⊗	⊗	⊗
Agency for International Development	•	⊗	⊗	⊗	⊗	⊗
Environmental Protection Agency	•	⊗	⊗	⊗	⊗	⊗
General Services Administration	•	⊗	⊗	⊗	⊗	⊗
National Aeronautics and Space Administration	•	⊗	⊗	⊗	⊗	•
National Science Foundation	•	⊗	⊗	⊗	⊗	⊗
Office of Personnel Management	•	⊗	⊗	⊗	⊗	⊗
Social Security Administration	•	⊗	⊗	⊗	⊗	⊗

• Yes; ⊗ No

Source: GAO analysis of survey results. | GAO-21-526.

Note: Agencies did not report using FRT for medical purposes in fiscal year 2020.

- *Digital access or cybersecurity.* Sixteen agencies reported using FRT for digital access or cybersecurity purposes.[12] Of these, 14 agencies authorized personnel to use FRT to unlock their agency- issued smartphones—the most common purpose of FRT reported by the agencies in our survey.[13] Two agencies—General Services Administration (GSA) and Social Security Administration (SSA)—reported conducting pilots that used agency employees to test FRT systems as a means to control access to certain government websites, such as GSA's login.gov.[14] Specifically, GSA and SSA used FRT to compare two images—a government photo identification and a live image of the individual—to verify the identity of an individual attempting to apply for an account. This FRT system may also conduct a check to detect if there is an attempt to subvert the FRT using a printed image or other non-live object.[15] However, agency officials said that this FRT would not be deployed until additional testing under a range of conditions is completed.
- *Domestic law enforcement.* Six agencies—Departments of Homeland Security (DHS), Justice (DOJ), Defense (DOD), Health and Human Services (HHS), the Interior, and the Treasury—reported using FRT to generate leads in criminal investigations, such as identifying a person of interest by comparing images of the person against

---

[12] GAO's cybersecurity work encompasses a broad range of issues assessing information security as a government-wide high-risk area, including protecting cyber critical infrastructure and protecting the privacy of personally identifiable information. This body of work also addresses agency compliance with federal cybersecurity requirements and includes assessments of security controls. As used in this report, cybersecurity refers to the general framework that includes digital access and other controls.

[13] The 14 agencies that reported using facial recognition to unlock smartphones or tablets are the Departments of Agriculture (USDA), Commerce, Homeland Security (DHS), Energy (DOE), Justice (DOJ), Health and Human Services (HHS), the Interior, the Treasury, Veterans Affairs (VA), the Environmental Protection Agency (EPA), the National Aeronautics and Space Administration (NASA), the National Science Foundation (NSF), the Office of Personnel Management (OPM), and the U.S. Agency for International Development (USAID). Many of these agencies reported that they did not require FRT when procuring their smartphones; rather it was a feature that was built into the smartphone. We did not inquire whether an agency monitored an employee's use of facial recognition to unlock their agency-issued smartphones or had policies related to its use. However, OPM reported that they no longer authorized personnel to use the feature.

[14] Login.gov is a publicly accessible website that verifies the identities of individuals seeking to access participating agencies' websites.

[15] This check detects whether a facial recognition system sensor is viewing data from a live subject as opposed to recorded data of a non-living object. For example, a picture or a 3D mask or a printed image that may be presented to try to fool the FRT system into false authentication.

databases of mugshots or from other law enforcement encounters.[16] For example, DOJ's Federal Bureau of Investigation used the Next Generation Identification Interstate Photo System to generate leads during investigations by comparing photos of unknown individuals suspected of criminal activity against a repository of photos of known individuals, including mugshots and other records. In addition, agencies may access commercially owned FRT systems as part of criminal investigations or to assist in identifying a missing person or victims of crimes, such as exploited children. For example, DHS, DOJ, HHS, and the Interior reported using Clearview AI, a commercially owned facial recognition system that compares a submitted photo against a database of publicly available images from open sources, such as social media, and returns matching images for review.

- *Physical security.* Five agencies—Department of Commerce, DOD, Department of Energy (DOE), DOJ, and HHS—reported using FRT to monitor or surveil locations to determine if an individual is present, such as someone from a watchlist, or to control access to a building or facility. For example, HHS reported that it used an FRT system (AnyVision) to monitor its facilities by searching live camera feeds in real-time for individuals on watchlists or suspected of criminal activity, which reduces the need for security guards to memorize these individuals' faces. This system automatically alerts personnel when an individual on a watchlist is present. In addition, DOJ reported using an FRT system to verify that personnel attempting entry into their on-site, secure network operations centers at federal prisons were authorized for entry.
- *Border and transportation security.* Two agencies—DHS and Department of State—reported using FRT systems to assist with identifying or verifying travelers within or seeking admission to the United States, identifying or verifying the identity of non-U.S.

[16] DOD's criminal investigative organizations (e.g., Naval Criminal Investigative Service and Air Force Office of Special Investigations) and other DOD law enforcement agencies or organizations (e.g., military police departments) can use FRT for domestic law enforcement, in addition to other purposes, when "that information logically relates to the detection, neutralization, or deterrence of criminal activity that affects DOD personnel, property, or mission." *See*, Department of Defense, DOD Instruction No. 5505.17, *Collection, Maintenance, Use, and Dissemination of Personally Identifiable Information and Law Enforcement Information by DOD Law Enforcement Activities*, (Washington, D.C., Dec. 19, 2012, rev. Nov. 29, 2016).

citizens already in the United States, and to research agency information about non-U.S. citizens seeking admission to the United States. For example, DHS's U.S. Customs and Border Protection used its Traveler Verification Service at ports of entry to assist with verifying travelers' identities. The Traveler Verification Service uses FRT to compare a photo taken of the traveler at a port of entry with existing photos in DHS holdings, which include photographs from U.S. passports, U.S. visas, and other travel documents, as well as photographs from previous DHS encounters.[17]

- *National security and defense.* Four agencies—DHS, DOD, DOJ, and State—reported using FRT for national security and defense purposes, including to identify individuals known or suspected to be terrorists, research derogatory information about a suspected threat actor, and monitor or surveil locations to search for a person of interest, such as a suspected terrorist. For example, the State Department reported using the Integrated Biometric System, which uses FRT to perform searches of visa and passport applicants' photos against terrorist watchlist photos. The Integrated Biometric System provides consular posts and passport agencies around the world with additional information to evaluate visa and passport applications to decrease the possibility that a terrorist would be able to fraudulently receive a U.S. visa or passport.
- *Other purposes.* Three agencies—DOD, DOJ, and the National Aeronautics and Space Administration (NASA)—reported using FRT for other purposes, including to verify the identities of individuals receiving identification cards and temporary badges, and demonstrate how FRT works in educational settings. For example, DOD tested a facial detection capability to support issuance of DOD identification cards within its Real-time Automated Personnel Identification System (RAPIDS). Specifically, RAPIDS uses facial detection when capturing the cardholder's picture during initial enrollment or identification card issuance or renewal to ensure that the size of the picture printed on the cards is consistent. Similarly, NASA's Johnson Space Center reported testing a prototype FRT

[17] See GAO-20-568 for more detail.

system to confirm an employee's identity by comparing a current camera image of the employee with a photo on file.[18]

Eighteen Agencies Reported Owning and Accessing Facial Recognition Technology

According to our analysis of survey responses, 18 agencies reported they owned FRT systems or accessed other entities' FRT systems in fiscal year 2020. Of these agencies, 17 owned or accessed federal FRT systems, three accessed FRT systems owned by state and local entities, and six accessed FRT systems owned by commercial vendors.

Seventeen Agencies Reported Owning or Accessing Federal FRT Systems

Seventeen agencies reported they owned or accessed 27 federal FRT systems.[19] Fourteen of these agencies owned smartphones that can be unlocked with facial recognition, and three of these agencies—U.S. Agency for International Development, EPA, and OPM—did not own or access any other FRT systems.[20] Nine agencies—Commerce, DOD, DOE, HHS, DHS, DOJ, State, GSA, and NASA—owned FRT systems other than smartphones, as shown in Table 3. Three of these agencies— DHS, DOD, and DOJ—owned 18 of the 27 federal FRT systems, in addition to owning smartphones. Finally, one agency—Treasury—did not own an FRT system, but accessed federal and commercially owned systems.[21]

---

[18] NASA reported that it conducted the experiment for a limited time only, and did not continue working on or using the prototype after fiscal year 2020.

[19] In some cases, agencies reported their systems as owned at the component-level (or below), and that access may be given to other components within the same agency. We report such cases as a single instance of ownership by the agency. Appendix II provides more detailed descriptions of the federal FRT systems, including how agencies use them, whether they were accessed by other agencies, and more information on the systems.

[20] We only counted smartphones that can be unlocked using facial recognition as a single FRT system, even though there are multiple vendors offering similar technologies on their smartphones. Fourteen agencies reported they owned smartphones that can be unlocked with facial recognition. These agencies include: USDA, Commerce, DOE, HHS, DOJ, DHS, Interior, Treasury, VA, USAID, EPA, NASA, NSF, and OPM.

[21] Treasury reported it accessed the General Services Administration's login.gov during testing of the FRT capability. Login.gov is a single-sign on mechanism that uses FRT to match applicants to their identification documents to access accounts on agency websites as mentioned earlier. Treasury also reported a third-party vendor performed facial recognition searches on its behalf in April 2020.

Table 3. Federally owned facial recognition technology (FRT) systems used in fiscal year 2020

Agency	Number of FRT Systems Owned	Accessed FRT Systems Owned by Other Agencies	Purposes
Department of Commerce	1 system	• None	• Physical security
Department of Defense	7 systems	• Department of Justice • Department of Homeland Security	• Physical security • Domestic law enforcement • National security and defense • Other
Department of Energy	1 system	• None	• Physical security
Department of Health and Human Services	3 systems	• None	• Physical security • Domestic law enforcement • Digital access or cyber security
Department of Homeland Security	4 systems	• Department of Defense • Department of Justice • Department of State	• Domestic law enforcement • Border and transportation security • National security and defense
Department of Justice	7 systems	• Department of Defense • Department of State	• Domestic law enforcement • Physical security • National security and defense • Other
Department of State	1 system	• Department of Defense	• Border and transportation security • National security and defense
General Services Administration	1 system	• None	• Digital access or cyber security
National Aeronautics and Space Administration	1 system	• None	• Other

Source: GAO analysis of survey results. | GAO-21-526.

Note: The table excludes agencies that only reported owning smartphones that can be unlocked with facial recognition. See appendix II for additional information about agency use of FRT systems.

Of the nine agencies that owned an FRT system other than smartphones, five agencies—Commerce, DOE, GSA, HHS, and NASA—reported they owned federal FRT systems and did not access other federal FRT systems. For example, HHS tested an FRT system that agency personnel could use to unlock their laptops. NASA tested another FRT system that verified an employee's identity by comparing a camera image with a photo on file if the employee forgot their badge. Commerce and DOE owned FRT systems that controlled personnel access to secure facilities.

For example, DOE's TacID Guard Dog performs facial matching and facial verification from live video to monitor the entry and exit of agency personnel at access points.

The other four agencies—DOD, DHS, DOJ, and State—reported that they owned an FRT system and also accessed FRT systems owned by other federal agencies. For example, the State Department's Bureau of Consular Affairs reported owning the Integrated Biometric System, which is also accessed by DHS and DOJ. State Department officials used this FRT system to verify a visa applicant's identity or determine whether an individual has previously applied for a visa under an alias. Similarly, DHS accessed the Integrated Biometric System to verify the identity of individuals applying for visas and immigration benefits, and to identify individuals being investigated for identity theft and benefits fraud. In addition, the State Department's Bureau of Diplomatic Security reported accessing DOD's Automated Biometric Identification System to verify the authenticity of an individual's travel documents, among other things.

Three Agencies Reported Accessing FRT Systems Owned by State and Local Entities

Three agencies—DOJ, DHS, and Interior—reported accessing one or more FRT systems owned by 29 states and seven localities for law enforcement purposes. Figure 3 shows the states and localities that own FRT systems accessed by these federal agencies.

DOJ and DHS reported direct and indirect access to FRT systems across states and localities.[22] For example, DOJ reported access to FRT systems through personnel in the FBI's Facial Analysis, Comparison, and Evaluation

[22] The term 'direct access' refers to cases where federal agency personnel can log into an FRT system and perform a facial recognition search. The term 'indirect access' refers to cases where federal agency personnel request that the owner of an FRT system conduct a facial recognition search.

(FACE) Services, which has memoranda of understanding with 21 states and two federal entities to access their FRT systems.

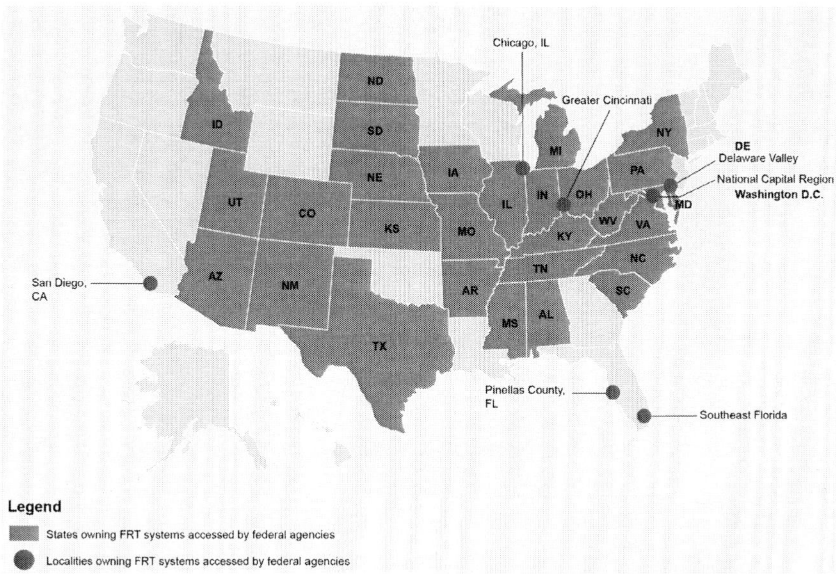

Source: GAO analysis of survey responses. | GAO-21-526.

Figure 3. States and Localities that Own Facial Recognition Technology (FRT) Systems Accessed by Federal Agencies in Fiscal Year 2020.

Specifically, FBI agents can submit a probe photo and request FACE Services examiners perform facial recognition searches for their investigations. FACE Services examiners have direct access to one state FRT system and indirect access to FRT systems owned by 20 other state entities.[23]

In addition, DHS reported it owns the Homeland Security Information Network (HSIN), which enables access to the Multi-State Facial Recognition Community of Interest. While HSIN is not an FRT system, it has a form for authorized users to request indirect facial recognition searches through state

[23] According to FBI officials, for direct access to the Maryland Department of Public Safety and Correctional Services' FRT system, examiners have completed related training requirements, so a memorandum of understanding is unnecessary. FACE Services has direct access to two federal FRT systems—the FBI's Next Generation Identification Interstate Photo System and the Department of State's visa holdings. The FBI can request the Department of State perform facial recognition searches on passport holdings on the FBI's behalf and return only a limited number of photos that are likely matches.

and local entities, such as fusion centers.[24] Through various memoranda of understanding, DHS personnel in U.S. Customs and Border Protection and U.S. Immigration and Customs Enforcement may submit photos for facial recognition searches to those entities. These searches assist DHS personnel with identifying individuals involved in identity theft and benefit fraud investigations, and targets connected to criminal investigations or to a known terrorist organization. In fiscal year 2020, DHS officials reported the agency made facial recognition search requests to 15 state and four local partner agencies, including state and local law enforcement fusion centers.

Finally, Interior reported it accessed the National Capital Region Facial Recognition Investigative Leads System (NCRFRILS). NCRFRILS is an FRT system that contains copies of information, including photos, from participating law enforcement agencies in the Washington, D.C. metro area. Interior's U.S. Park Police reported that in July 2020 it requested the Maryland National Capital Park Police perform a facial recognition search of NCRFRILS on its behalf to generate investigative leads.

Six Agencies Accessed Commercial FRT Systems

Six agencies—HHS, DHS, Interior, DOJ, Treasury, and SSA—reported accessing eight FRT systems owned by commercial vendors.

As shown in Table 4, four agencies—HHS, DHS, Interior, and DOJ—accessed Clearview AI, which conducts facial recognition searches using publicly available images.[25] For example, HHS's Office of the Inspector General reported it began a pilot of Clearview AI in September 2020 to assist with identifying subjects of a criminal investigation. Similarly, DHS reported the U.S. Secret Service and U.S. Immigration and Customs Enforcement piloted Clearview AI in fiscal year 2020, to identify individuals in federal criminal investigations, and to identify perpetrators and victims in domestic and international child exploitation cases, respectively. DHS also reported that, since June 2019, U.S. Customs and Border Protection agents had direct, albeit temporary, access to a Clearview AI through their participation with the New York State Intelligence Center. These agents used Clearview AI to

---

[24] HSIN is for trusted sharing of Sensitive but Unclassified information—such as that related to law enforcement or homeland security—between federal, state, local, territorial, tribal, international, and private sector partners. Generally, fusion centers are collaborative information-sharing efforts to detect, prevent, investigate, and respond to potential criminal activity, including terrorism.

[25] Interior's U.S. Park Police reported that it stopped using Clearview AI in June 2020 after conducting a pilot test in April 2020.

identify criminals, as well as subjects who have been arrested previously, were deported, and attempted to re-enter the United States at the border.

Table 4. Commercially owned facial recognition technology (FRT) systems accessed by federal agencies in fiscal year 2020

Agency	Number of FRT Systems Accessed	Commercial Vendor	Purposes
Department of Health and Human Services	1 system	• Clearview AI	• Domestic law enforcement
Department of Homeland Security	2 systems	• Clearview AI • Vigilant Solutions	• Domestic law enforcement • Border and transportation security • National security and defense
Department of the Interior	1 system	• Clearview AI	• Domestic law enforcement
Department of Justice	2 systems	• Clearview AI • Vigilant Solutions	• Domestic law enforcement
Department of the Treasury	1 system	• Other	• Domestic law enforcement
Social Security Administration	1 system	• Acuant FaceID	• Digital access or cybersecurity

Source: GAO analysis of survey results. | GAO-21-526.
Note: See appendix II for additional information about agency use of FRT systems.

Ten Agencies Reported Conducting or Supporting FRT-related Research and Development

Based on our analysis, 10 of the 24 agencies surveyed—Commerce, DHS, DOD, DOJ, the Department of Transportation (DOT), HHS, VA, NASA, the National Science Foundation (NSF), and State—conducted or supported research and development (R&D) for FRT in fiscal year 2020.[26] Of these agencies, DOT did not report any other FRT-related activities.

Four agencies—DHS, DOD, DOJ, and State—generally focused their R&D on agency-specific needs, such as to develop new applications or improve existing capabilities. Examples from each of these agencies include:

[26] In our questionnaire, we asked, "In fiscal year 2020, did your agency conduct research and development (R&D) involving facial recognition technologies?"

- DHS reported sponsoring Biometric Technology Rallies, which are ongoing industry challenges to develop innovative solutions for biometric collection and matching, including facial recognition. For example, the 2020 Rally focused on the ability of FRT systems to reliably collect or match images of individuals wearing masks. DHS intended this research to improve the technology's ability to recognize individuals without requiring them to remove their protective equipment.
- DOD reported researching new capabilities for RAPIDS, which would support identity verification during online identification card renewal and PIN reset requests.
- DOJ reported conducting applied research on the relationship between skin tone and false match rates in facial recognition algorithms, the capabilities and limitations of current synthetic face detection, such as deepfakes, and the development of software to detect synthetic faces.[27] DOJ also explored the potential benefits of combining FRT systems with trained forensic examiners to achieve better matching performance than by the technology or by humans alone.
- The State Department reported conducting research and development and contributing to international image standards for travel documents. For example, State conducted research on morphing detection and the impact of aging on the accuracy of facial recognition algorithms, such as for children's passport photos.[28] State is also developing the ability to match images of individuals against passport images and a repository of known or suspected terrorists using the Personal Identification Secure Comparison and Evaluation System.[29]

Two agencies—Commerce and NSF—conducted or supported FRT-related research more broadly, including for commercial vendors and other

---

[27] For more information on deepfakes, see GAO, *Science & Tech Spotlight: Deepfakes*, GAO-20-379SP (Washington, D.C.: Feb. 20, 2020).

[28] Face morphing is when, for example, two images of different individuals are combined so that the resulting image could be used as identification for both of them, such as on a passport. Morphing detection attempts to identify these images.

[29] The State Department's Bureau of Counterterrorism offers the Personal Identification Secure Comparison and Evaluation System, or PISCES, for border management under the Terrorist Interdiction Program to foreign partners, and incorporates enhanced screening technologies to ensure those partners are able to protect themselves from attempts by terrorists to enter, transit, or depart their country.

agencies. For example, Commerce reported that its National Institute of Standards and Technology performed research to support the development of standards and methods in performance measurement, image quality, testing and evaluating technologies, and interoperability for facial recognition technology. The National Institute of Standards and Technology also conducted research through the Face Recognition Vendor Test program, which most recently released reports quantifying facial recognition accuracy with facemasks using post-COVID-19 algorithms and across demographic effects.[30]

The NSF reported that it awards grants to universities and others to conduct research on facial recognition. Specifically, NSF's Directorate for Computer and Information Science and Engineering supported FRT- related research, including a project assessing how to prevent identifying an individual from facial images used in research, such as recordings of a driver's face during driver behavior studies. Furthermore, NSF supported a program called the Center for Identification Technology Research, in which university partners work with government and industry stakeholders on biometrics, including research on FRT.[31]

Finally, four agencies—DOT, HHS, the Department of Veterans Affairs (VA), and NASA—reported using FRT as a tool to conduct other research. For example, DOT reported that the Federal Railroad Administration used eye tracking to study alertness in train operators.

Like DOT, NASA also reported that it used eye tracking to conduct human factors research.

In addition, VA reported it used eye tracking as part of a clinical research program that treats post-traumatic stress disorder in veterans. Specifically, the eye tracking system evaluates pupil response to evaluate impairment. Similarly, HHS's National Institutes of Health awarded grants for research that use eye tracking as a tool for clinical research. For example, characterizing how children with and without autism spectrum disorder visually followed conversations in videos, among other areas of research.

[30] DHS and DOJ have interagency agreements with the National Institute of Standards and Technology for related FRT research and evaluation.

[31] DHS also sponsored the Center for Identification Technology Research.

Four Agencies Reported Other FRT-Related Activities with Nonfederal Entities

According to our analysis of survey responses, four agencies—DHS, DOJ, VA, and State—reported FRT-related activities with nonfederal entities in fiscal year 2020. Specifically, these agencies reported entering into transactions that enabled nonfederal entities to obtain their own FRT.[32] These transactions included agreements with foreign governments and commercial entities, grants, and medical equipment purchases. In addition, DHS was the only agency that reported it regulated an airline's use of FRT in fiscal year 2020.[33]

Transactions with Nonfederal Entities to Obtain FRT

Agreements

Two agencies—DHS and State—reported entering into FRT-related agreements in fiscal year 2020 with foreign governments, and DHS reported having FRT-related agreements with commercial entities. Specifically, DHS entered into agreements, called project arrangements, with two foreign governments—Australia and the United Kingdom—related to the assessment of facial recognition software. The State Department reported transactions with two foreign governments. Specifically, State contracted trainers to instruct Mexican government personnel on how to use previously donated FRT equipment, and it purchased FRT equipment to donate to the Guatemalan government. DHS also had cooperative research and development agreements with two commercial entities focused on making digital identity credentials (e.g., digital driver's licenses) interoperable with airport checkpoint security systems.

---

[32] In our questionnaire we asked, "In fiscal year 2020, did your agency enter into transactions to enable a nonfederal entity to obtain facial recognition technology for their own uses? In other words, the support (financial or in-kind) would enable nonfederal entities to develop, purchase, or use facial recognition technology for their own uses—not for your agency's use." For the purposes of this questionnaire and report, "transactions" refers to an agency that awarded grants; entered into contracts, leases, or cooperative agreements; provided direct loans or loan guarantees; or entered into any other transactions with nonfederal entities using other transactional authority.

[33] The Transportation Security Administration within the Department of Homeland Security regulated an airline's use of FRT in fiscal year 2020.

Contract
DHS's U.S. Immigration and Customs Enforcement reported that it awarded a contract to the Lehigh County, Pennsylvania District Attorney's Office to enhance U.S. Immigration and Customs Enforcement's future access to the Gang Intelligence Application, which is a database of transnational gang members and associated information.

Grants
Two agencies—DHS and DOJ—reported awarding FRT-related grants to nonfederal entities. Specifically, DHS's Federal Emergency Management Agency awarded preparedness grants to sustain or build facial recognition and other capabilities, such as intelligence sharing, among state and local law enforcement, emergency management, and other local entities.[34]

DOJ reported it awarded a grant to the Police Foundation for the development of techniques to automate analysis of body-worn camera audio and video data of police and community interactions. For example, these techniques could allow for an evaluation of officers' adherence to principles of procedural justice.

Medical Equipment
VA reported two FRT-related transactions for nonfederal entities. Specifically, VA purchased two types of eye tracking equipment for veterans. According to the VA, speech-impaired veterans were provided a prosthetic device that tracks eye movements to assist their use of a computer or tablet for communication.

Regulating Nonfederal Entities' Use of FRT
Of the 24 agencies in our survey, one agency—DHS—reported regulating the use of FRT by other entities in fiscal year 2020.[35] Specifically, the TSA issued

[34] DHS officials provided a list of the following preparedness grant programs that included "FRT" or "facial recognition" in the project description for awards in fiscal year 2020: Nonprofit Security Grant Program-Urban Area, State Homeland Security Program, and Urban Area Security Initiative.

[35] In our questionnaire, we asked, "In fiscal year 2020, did your agency engage in any regulatory functions over nonfederal entities that use facial recognition technology?" Regulated refers to using regulatory authority over a nonfederal entity to regulate that entity's use of its own FRT. For the purposes of this questionnaire and report, we defined "regulated" as regulatory functions in which the agency engaged, including but not limited to, investigatory and inspections activities, taking enforcement actions, prescribing requirements or guidance, conducting oversight, and maintaining performance standards.

security program amendments to aircraft operators that permit the use of FRT to identify passengers checking baggage for transportation on flights, in lieu of the standard passenger identification measures in the applicable TSA-issued security program.[36]

Ten Agencies Plan to Expand Use of FRT, Mostly through Use of New FRT Systems

According to our analysis of survey responses, 10 of the 24 agencies surveyed—USDA, Commerce, DOD, HHS, DHS, Interior, DOJ, State, Treasury, and VA—plan to expand their use of FRT systems in one or more ways through fiscal year 2023.[37] We categorized plans to expand FRT use in three ways: (1) using new FRT systems, (2) evaluating existing FRT systems (e.g., pilot testing), and (3) upgrading existing FRT systems.[38] See Table 5 for additional information.

---

[36] The TSA responded that it has broad authority to ensure transportation security (See, for example, 49 U.S.C. §§ 114, 44901, 44903), including research and development of new technologies (49 U.S.C. § 44912). Specific to biometrics, Congress authorized the TSA to use "voice stress analysis, biometric, or other technologies to prevent a person who might pose a danger to air safety or security from boarding the aircraft of an air carrier or foreign air carrier in air transportation or intrastate air transportation." (Aviation and Transportation Security Act of 2001 (Pub. L. No. 107-71, §109(a)(7), 115 Stat. 597, (2001)), codified at 49 U.S.C. § 114 note.

[37] In our questionnaire, we asked, "Does your agency have plans to begin using facial recognition technology (including upgrading a system to include facial recognition) for internal agency use between fiscal year 2020 and fiscal year 2023? In other words, your agency has not yet begun to use the facial recognition technology, but it has taken steps to begin by fiscal year 2023." For the purposes of this questionnaire and report, by "plans," we meant that the agency has initiated a process to use facial recognition technology, which could include an ongoing acquisition process, a contract with a vendor or another agency, a memorandum of understanding, or a budget request. It did not include hypothetical or exploratory conversations about potential uses of facial recognition technology within the agency. Furthermore, we did not ask agencies to confirm if they were planning to continue using their existing FRT systems beyond fiscal year2020. However, we included existing FRT systems in the planned use section when an agency reported plans to change the way it will use the FRT system from fiscal year 2020 through fiscal year 2023.

[38] For the purposes of this report, "upgrades" refers to cases where agencies reported adding FRT capabilities to existing systems or to enhance system processes and updating a previously deactivated FRT system. It is not typical system maintenance.

Table 5. Federal agencies that reported plans to expand their use of facial recognition technology (FRT) systems, through fiscal year 2023

Federal Agency	Plan to use new FRT systems	Plan to evaluate FRT systems	Plan to upgrade FRT systems or capabilities
Department of Agriculture	•	⊗	⊗
Department of Commerce	•	⊗	⊗
Department of Defense	•	•	⊗
Department of Health and Human Services	•	⊗	⊗
Department of Homeland Security	•	•	•
Department of the Interior	•	⊗	⊗
Department of Justice	•	⊗	⊗
Department of State	•	⊗	⊗
Department of the Treasury	•	•	⊗
Department of Veterans Affairs	•	⊗	⊗

• Yes

⊗ No

Source: GAO analysis of survey results. | GAO-21-526.

Using New FRT Systems

Ten agencies—USDA, Commerce, DOD, HHS, DHS, Interior, DOJ, State, Treasury, and VA—reported plans to use 17 new FRT systems through fiscal year 2023 as shown in Table 6.[39] New FRT systems refers to systems that are new to federal agencies, including newly-developed FRT systems and commercial-off-the-shelf systems, and new access to existing FRT systems that agencies did not report using in fiscal year 2020.

According to our analysis of survey results, agencies reported that 13 of the 17 new FRT systems will be owned by federal agencies, and two by local governments. Two agencies reported they plan to access Clearview AI, a commercial system, for the first time.[40]

[39] See appendix II for additional information about agency use of FRT systems.

[40] Multiple agencies already accessed Clearview AI in fiscal year 2020 (noted above).

Table 6. New facial recognition technology (FRT) systems federal agencies reported they plan to use, through fiscal year 2023

Agency	Planned Number of Newly Owned and Accessed FRT Systems	Planned Access to FRT Systems Owned by Other Agencies	Purposes
Department of Agriculture	2 systems	• None	• Physical security • Domestic law enforcement
Department of Commerce	1 system	• None	• Physical security
Department of Defense	2 systems	• None	• Physical security • Domestic law enforcement • National security and defense
Department of Health and Human Services	1 system	• None	• Domestic law enforcement
Department of Homeland Security	2 systems	• None	• Domestic law enforcement • National security and defense
Department of the Interior	2 systems	• None	• Domestic law enforcement
Department of Justice	2 systems	• None	• Physical security • Border and transportation security
Department of State	1 system	• None	• Border and transportation security
Department of the Treasury	2 systems	• Department of Defense • Department of Homeland Security • Department of Justice	• Domestic law enforcement
Department of Veterans Affairs	2 systems	• None	• Physical security • Domestic law enforcement

Source: GAO analysis of survey results. | GAO-21-526.

Note: See appendix II for additional information about agency use of FRT systems.

Use of New Federal FRT Systems

Nine agencies—USDA, Commerce, DOD, HHS, DHS, DOJ, State, Treasury, and VA—plan to use new federal FRT systems. For example, the U.S. Treasury Inspector General for Tax Administration reported that it purchased an FRT system that can identify facial images of persons of interest who may be involved in criminal activity across multiple investigations in December 2020. The FRT system searches images in an online storage locker, which contains evidence such as photos from seized mobile devices, and will notify investigators of potential matches of individuals linked to other investigations.

The State Department reported"plans for a p"lot in late 2021, using FRT developed for the Personal Identification Secure Comparison and Evaluation System (PISCES) border management system. State plans to screen individuals against passport images and a repository of suspicious individuals, such as known and suspected terrorists attempting to travel through partner countries.

New Access to Existing FRT Systems

Three agencies—DOD, Interior, and Treasury—reported plans to access existing FRT systems.[41] For example, DOD's U.S. Air Force Office of Special Investigations reported it began an operational pilot using Clearview AI in June 2020, which supports the agency's counterterrorism, counterintelligence, and criminal investigations. The agency reported it already collects facial images with mobile devices to search national databases and plans to enhance searches by accessing Clearview AI's large repository of facial images from open sources to search for matches.

Evaluating Existing Systems

Three agencies—DHS, DOD, and Treasury—reported plans to conduct new pilot tests or continue evaluating existing FRT systems. Of the four FRT systems these agencies plan to evaluate, federal agencies own three systems, and a commercial vendor owns the other system.

For example, DHS reported plans to initiate a new pilot and continue an ongoing pilot of an existing FRT system. As of March 2021, the TSA is collaborating with U.S. Customs and Border Protection and a commercial airline at the Detroit Metropolitan Wayne County Airport to evaluate the use of biometric technology, including facial recognition, to automate the identity verification process at TSA checkpoints and streamline traveler experience.

In addition, DOD and Treasury reported plans to conduct new pilots of existing FRT systems. For example, DOD plans to conduct a pilot in late fiscal

---

[41] These systems are already used by other agencies.

year 2021 of an FRT enhancement to an electronic physical access control FRT system, called Automated Installation Entry, to improve processing and minimize security risks.[42] DOD personnel that volunteer for the new pilot will proceed to the enhanced access control points, which will match their faces against a database of DOD participants.

Upgrading Existing Systems

DHS reported plans to upgrade an existing FRT system and capabilities through fiscal year 2023. In December 2021, DHS plans to replace IDENT, which is its current system for processing and storing biometric data, with the Homeland Advanced Recognition Technology system.

Initially, the Homeland Advanced Recognition Technology system will replace IDENT's current capabilities, to include FRT, with DHS planning additional capabilities in subsequent years.

Agency Comments

We provided a draft of this report to the 24 CFO Act agencies for their review and comment. We received written comments from USAID and SSA that are reprinted in appendices III and IV, respectively. USAID in its written comments did not comment on the content of the report. SSA in its written comments noted that the report was accurate with respect to their experience with facial recognition technologies. We received technical comments from six agencies, which we incorporated as appropriate. We did not receive comments from the Department of the Treasury's offices, but received technical comments from some of its bureaus and components. The remaining 15 agencies informed us that they had no comments.

Candice N. Wright
Director
Science, Technology Assessment, and Analytics

[42] The U.S. Army purchased the FRT system in fiscal year 2014 to verify individuals seeking access to military installations.

Gretta L. Goodwin
Director
Homeland Security and Justice

Appendix I : Objectives, Scope, and Methodology

This chapter identifies and describes (1) how agencies used facial recognition technologies (FRT) in fiscal year 2020, including any FRT- related research and development activities and interactions with nonfederal entities, and (2) how agencies plan to expand their use of FRT through fiscal year 2023.

To obtain the information needed for both objectives, we conducted a survey of the 24 agencies listed in the Chief Financial Officers (CFO) Act of 1990, as amended.[43] These departments and independent agencies (hereafter referred to as agencies) are as follows:

- Department of Agriculture
- Department of Commerce
- Department of Defense
- Department of Education
- Department of Energy
- Department of Health and Human Services
- Department of Homeland Security
- Department of Housing and Urban Development
- Department of the Interior
- Department of Justice
- Department of Labor
- Department of State
- Department of Transportation
- Department of the Treasury
- Department of Veterans Affairs
- Agency for International Development
- Environmental Protection Agency
- General Services Administration

[43] The 24 agencies are those identified in the Chief Financial Officers Act of 1990, as amended (31 U.S.C. § 901(b)). This Act does not include many independent agencies and commissions, such as the Office of the Director of National Intelligence or the Central Intelligence Agency, so they were not included in our survey.

- National Aeronautics and Space Administration
- National Science Foundation
- Nuclear Regulatory Commission
- Office of Personnel Management
- Small Business Administration
- Social Security Administration

We administered a questionnaire by email to each of the 24 CFO Act agencies from October 2020 through January 2021.[44] During survey administration, we asked agencies to provide responses for all their components, bureaus, and offices in a consolidated response. We received responses from all 24 agencies.[45]

To develop the questionnaire, we used information from prior GAO reports and early interviews with the agencies to determine the areas of inquiry. For the purposes of our survey and report, we defined facial recognition technology as systems, components, or modules of systems, software applications, or devices with automated facial recognition capabilities, such as face recognition algorithm, hardware, or software. Facial recognition generally refers to facial matching, which includes both verification (one-to-one matching)—to automatically confirm whether a facial image in one photo matches a facial image in a different photo— and identification (one-to-many matching)—to automatically determine whether a facial image has any match in a database or gallery of photos.

Though they are generally considered distinct technologies, we also considered facial analysis—identifying attributes about a person based on their face, such as sex, age, or emotion—and facial detection— determining if a photo or video contains a face— to be facial recognition technologies.

We also determined that facial recognition technology could be used for a variety of applications, such as verifying the identity claimed by an individual or controlling access to buildings or computers. We grouped these applications into seven purposes based on prior GAO reports, such as those on law enforcement, transportation security, and commercial uses of facial recognition, and a review of relevant literature.[46] We asked agencies to include

[44] These dates cover the initial response for all 24 agencies. It does not include updated or additional responses received because of follow-up activities.

[45] We also asked agencies to include the 24 Offices of Inspectors General and the Treasury Inspector General for Tax Administration in their response.

[46] GAO, *Face Recognition Technology: FBI Should Better Ensure Privacy and Accuracy*, GAO-16-267 (Washington, D.C.: May 16, 2016); *Facial Recognition: CBP and TSA are Taking*

all uses of facial recognition technology and identify which purpose(s) described their uses. The purposes are: (1) physical security; (2) digital access or cybersecurity; (3) domestic law enforcement; (4) border and transportation security; (5) national security and defense; (6) medical assessment; and (7) other purposes not already listed.

Finally, we met with both agency liaisons, who would be administering the survey to their respective agencies, and subject matter experts, who helped ensure respondents clearly understood specific questions.

We used a questionnaire with two sections: (1) a Main Questionnaire, and five Attachments that followed up on positive responses in the Main Questionnaire. For agencies that had activities or planned activities related to facial recognition technology in the Main Questionnaire, we asked that, as appropriate, the agencies complete an Attachment that covered the following areas:

1. owned or accessed, including tested, facial recognition technology in fiscal year 2020;
2. plans to own or access, including testing, facial recognition technology through fiscal year 2023;
3. research and development conducted or supported in fiscal year 2020;
4. any transactions the agency entered into with nonfederal entities for that entity to obtain facial recognition technology in fiscal year 2020; and
5. any regulation of a nonfederal entity's use of its own facial recognition technology.[47]

We used these timeframes because fiscal year 2020 was the most recent fiscal year for which information was available when we issued our questionnaire and through fiscal year 2023 for planned systems, because

Steps to Implement Programs, but CBP Should Address Privacy and System Performance Issues, GAO-20-568 (Washington, D.C.: Sept. 2, 2020); *Facial Recognition Technology: Commercial Uses, Privacy Issues, and Applicable Federal Law*, GAO-15-621 (Washington, D.C.: July 30, 2015); and *Facial Recognition Technology: Privacy and Accuracy Issues Related to Commercial Uses*, GAO-20-522 (Washington, D.C.: July 13, 2020).

[47] Regulated refers to using regulatory authority over a nonfederal entity to regulate that entity's use of its own FRT. For the purposes of our questionnaire, we defined "regulated" as regulatory functions in which the agency engaged, including, but not limited to, investigatory and inspections activities, taking enforcement actions, prescribing requirements or guidance, conducting oversight, and maintaining performance standards.

agencies were most likely to have information covering this timeframe, such as in strategic plans.

Each of the Attachments had detailed questions, such as a description of the activity, how the agency used the technology, including the purpose(s) and obligations related to the activity. For how agencies used facial recognition technology, we asked whether the federal agency owned or accessed it directly (e.g., logging into a system) or via a third party (e.g., asking another entity to run the search on its behalf). For example, we asked the name of the technology and the entity that owned it66ncludeing how best to describe that entity (e.g., federal, state, tribal, or local). For obligations, we asked the agency to provide the amount obligated in fiscal year 2020 and whether it was disbursed. If the agency could not provide an obligated amount we asked for a range or the reason why an agency did not know (e.g., facial recognition was a small part of a larger, biometric system and was not specifically tracked). Finally, we also asked if agencies had classified systems.[48]

After determining the areas of inquiry, we conducted pretests with five agencies to test the questionnaire's applicability to all agencies and a variety of facial recognition technology uses, and revised the questionnaire based on those pretests. We selected agencies for pretests based on information provided during initial meetings with agency officials to capture a variety of facial recognition technologies and purposes, in order to test different parts of our questionnaire. For example, National Aeronautics and Space Administration officials told us they used facial recognition technology for research and development, so we focused the pretest on the research and development questions with that agency. On the other hand, Department of Defense officials were able to pretest the majority of the questions, so we met with them multiple times.

Once the design process was completed, we administered the questionnaire by email to agency liaisons, or their designees, for an agency-level response about facial recognition technology use. To do this, we instructed the liaisons to provide the agency-level response in the Main Questionnaire. We asked the liaisons to disseminate the Attachments to relevant and knowledgeable facial recognition technology subject matter experts at the agency, component, bureau, or office levels, which the liaisons would consolidate into the Main Questionnaire and return it and all the

[48] This report only discusses unclassified systems.

Attachments to us.[49] We provided detailed instructions in writing and in the questionnaire itself. We also followed up by email and phone, when appropriate, to ensure that the agency liaisons received the questionnaire and to ask if they had any questions about it. Once the liaison determined their agency's survey responses were complete, we asked the liaison to total the Attachments and enter the total number of completed Attachments on the Main Questionnaire to ensure we received all of the expected agency Attachments in their survey responses.

When agencies submitted their survey responses, we conducted an initial review for completeness, inconsistencies, or logical errors within the responses. We asked agencies to re-submit or clarify responses if necessary.

Because we surveyed and obtained responses from all 24 agencies in the population defined by our scope, the summary results describing this group are not subject to errors from sampling and nonresponse. However, the practical difficulties of conducting any survey may introduce other errors, such as:

- Difficulties in how a particular question is interpreted by respondents. For example, some agencies were not sure which Attachment to fill out or completed one for an activity that did not meet our criteria, so we followed up with them to determine which was the most appropriate in some cases. Furthermore, as part of our analysis of agency responses, we determined that some information could have been included correctly in more than one Attachment. In one case, for agencies that reported using facial recognition technology, such as eye tracking, as part of research unrelated to facial recognition for other purposes, we determined that those technologies would only be reported as research and development activities, even if some agencies reported it as a system "used." In another case, for agencies that filled out the regulatory attachment, we excluded responses that did not meet both of our criteria: (1) that an agency had authority over a nonfederal entity's use of facial recognition technology and (2) actually used that authority to regulate the nonfederal entity's use of facial recognition technology.

[49] For example, the Departments of the Treasury and Defense did not consolidate their responses. We informed them that we would take the responses provided by these agencies and consolidate them on behalf of the agency and provide them with an opportunity to review this information in the report.

- Sources of information that are available to respondents differ across agencies. For example, one agency expressed concern to us about their ability to provide a comprehensive response that included every accessed facial recognition technology because they did not track this information. We asked that agencies provide what they could, but to focus their efforts on facial recognition technology access that had a memorandum of agreement or understanding. Furthermore, we relied on the agency liaisons to provide Attachments to subject matter experts on facial recognition technology use within their agencies. We provided suggestions to these liaisons of possible areas where facial recognition technology could be used, such as security and information technology offices, to ensure that the agency-level responses were comprehensive.

To help corroborate the information agencies provided in the questionnaire, we conducted a search of government contracting information and reviewed information provided for prior reports. Specifically, we conducted a search of several terms, such as "facial recognition," "FRT," and known vendor names, in the Federal Procurement Data System Next Generation and Grants.gov databases. We used two analysts to independently review and determine if a result was related to FRT in our scope, such as excluding non-human related results and other technologies. We used this list to review each agency's responses for completeness. When we discovered discrepancies, we followed up with the agency as appropriate to change their response or fill out new Attachments from the questionnaire.

Furthermore, we reviewed responses from a survey of federal law enforcement use of facial recognition technology to determine if there were inconsistencies in responses to our survey.[50] If there were, we requested clarification of the agency-provided information, such as system descriptions, in interviews and multiple rounds of follow up (i.e., information requests), to those agencies. For example, if an agency filled out an Attachment, but indicated "no" on the Main Questionnaire for the applicable question related to the Attachment,

---

[50] For more information about the survey of federal law enforcement's use of facial recognition technology, see GAO, *Facial Recognition Technology: Federal Law Enforcement Agencies Should Better Assess Privacy and Other Risks*, GAO-21-518 (Washington, D.C., June 3, 2021).

we determined that was an incorrect response on the Main Questionnaire.

- How we processed and analyzed the responses we received can influence the accuracy of the survey results. For example, we consolidated some agency responses, such as the Department of Defense. We independently verified the consolidated information internally and presented it to those agencies prior to issuing the report.

We took steps in the development of the questionnaire, such as pretesting, data collection, and data analysis, including multiple rounds of follow up through interviews and information requests as noted above to minimize these potential errors and to help ensure the accuracy of the answers obtained. Based on these quality assurance and control actions we determined that for the purposes of this chapter, the information provided is an accurate and valid representation of the extent of facial recognition use across the 24 CFO Act agencies.

We conducted this performance audit from April 2020 through August 2021 in accordance with generally accepted government auditing standards. Those standards require that we plan and perform the audit to obtain sufficient, appropriate evidence to provide a reasonable basis for our findings and conclusions based on our audit objectives. We believe that the evidence obtained provides a reasonable basis for our findings and conclusions on our audit objectives.

Appendix II : Summaries of Selected Federal Agencies' Facial Recognition Technology Activities

This appendix provides summaries of those Chief Financial Officers Act agencies that reported facial recognition technology (FRT) activities in fiscal year 2020, and planned FRT systems through fiscal year 2023. We asked the following questions of each agency, which correspond to the responses provided in each agency's summary:

1. "At any point in fiscal year 2020, did your agency use facial recognition technology for any of the following purposes? By 'use,' we mean your agency: (1) owned and/or operated facial recognition technology for internal agency purposes, (2) accessed another federal

or nonfederal entity's (including local government or private company) facial recognition technology under an agreement or arrangement as part of an agency program or activity, or (3) tested facial recognition technology as part of a pilot, proof of concept, trial, or evaluation for potential agency use."

2. "Does your agency have plans to begin using facial recognition technology (including upgrading a system to include facial recognition) for internal agency use between fiscal year 2020 and fiscal year 2023? In other words, your agency has not yet begun to use the facial recognition technology, but it has taken steps to begin by fiscal year 2023. By 'plans,' we mean your agency has initiated a process to use facial recognition technology, which could include an ongoing acquisition process, a contract with a vendor or another agency, a memorandum of understanding, or a budget request. Do not include hypothetical or exploratory conversations about potential uses of facial recognition technology within your agency."
3. "In fiscal year 2020, did your agency conduct research and development (R&D) involving facial recognition technologies? This could include funding another entity to conduct R&D on your agency's behalf. R&D includes basic research, applied research, or experimental development (technology readiness levels 1-6). R&D could include developing facial recognition algorithms or evaluating existing algorithms."
4. "In fiscal year 2020, did your agency enter into transactions to enable a nonfederal entity to obtain facial recognition technology for their own uses? In other words, the support (financial or in-kind) would enable nonfederal entities to develop, purchase, or use facial recognition technology for their own uses—not for your agency's use. By 'transactions,' we mean your agency awarded grants, entered into contracts, leases, or cooperative agreements, provided direct loans or loan guarantees, or entered into any other transactions using other transactional authority."
5. "In fiscal year 2020, did your agency engage in any regulatory functions over nonfederal entities that use facial recognition technology? For our purposes, 'regulatory functions' includes, but is not limited to, investigatory and inspections activities, taking enforcement actions, prescribing requirements or guidance, conducting oversight, and maintaining performance standards."

Specifically, we provide summaries for the following 16 agencies:

- Department of Agriculture
- Department of Commerce
- Department of Defense
- Department of Energy
- Department of Health and Human Services
- Department of Homeland Security
- Department of the Interior
- Department of Justice
- Department of State
- Department of Transportation
- Department of the Treasury
- Department of Veterans Affairs
- General Services Administration
- National Aeronautics and Space Administration
- National Science Foundation
- Social Security Administration

The information in these summaries is from survey responses or requests for more information from the agencies. We present details on FRT systems reported by agencies that own their FRT systems or access the FRT systems of other government entities, including federal and nonfederal governments, and commercial facial recognition service providers. Agencies can have direct access to an FRT system, such as by logging into the system, or indirect access, such as by requesting a state government (i.e., a third party) run a facial recognition search on behalf of the agency.

We also present obligations for FRT systems as applicable. Each agency provided information on obligations in its survey response. However, we are presenting obligations data for informational purposes only, because we did not corroborate it through other means, such as document requests, and some agencies reported facial recognition obligations as part of a larger biometric system. We do not include information on obligations related to unlocking smartphones or tablets because agencies reported obligations of (1) none or free, because the smartphones were included in the service contract, (2) the cost of individual devices, or (3) the total of all purchased smartphones.[51] The facial recognition feature of the smartphones was included with the phones, so there is no specific obligation.

The Department of Education, the Department of Housing and Urban Development, the Nuclear Regulatory Commission, and the Small Business Administration do not have summaries because they reported they had no FRT activities in fiscal year 2020 and no plans to have FRT activities through fiscal year 2023. The Department of Labor, the U.S. Agency for International Development, the Environmental Protection Agency, and the Office of Personnel Management do not have summaries because they reported that they only use facial recognition to unlock smartphones or tablets.

USDA

The U.S. Department of Agriculture (USDA) reported that it did not have facial recognition technology (FRT) activities in fiscal year 2020. However, it reported plans to use a new FRT system through fiscal year 2023.

Use of FRT Systems in Fiscal Year 2020

None reported.

FRT System Obligations in Fiscal Year 2020

None reported.

Research & Development in Fiscal Year 2020

None reported.

Transactions with Nonfederal Entities in Fiscal Year 2020

None reported.

Regulation of Nonfederal Entities' Use of FRT in Fiscal Year 2020

None reported.

Planned Use of FRT Systems through Fiscal Year 2023

USDA reported it plans to use two other FRT systems through fiscal year 2023, both of them for physical security purposes and one of them also for domestic law enforcement purposes.

---

[51] The 14 agencies that reported using facial recognition to unlock smartphones or tablets are the Departments of Agriculture, Commerce, Homeland Security, Energy, Justice, Health and Human Services, the Interior, the Treasury, Veterans Affairs, the Environmental Protection Agency, the National Aeronautics and Space Administration, the National Science Foundation, the Office of Personnel Management, and the U.S. Agency for International Development.

U.S. Department of Agriculture (USDA) Plans for Facial Recognition Technology (FRT) Systems through Fiscal Year 2023		
FRT system	**Description of system use**	**Planned new use status**
New federal FRT systems		
Agricultural Research Service IDEMIA VisionPass	IDEMIA VisionPass is a facial recognition device that will be used to verify the identity of personnel for access within secure areas of a facility.	USDA plans to use the technology beginning in fiscal year 2022.
Office of Safety, Security, and Protection Avigilon Control Center	The Avigilon Control Center facial recognition software will assist with accelerating response times by identifying individuals of interest based on secure watch lists.	USDA plans for the facial recognition software to be operational by fiscal year 2022, if funding is approved.

Source: GAO analysis of survey results. | GAO-21-526

Components with No FRT-Related Activities in Fiscal Year 2020 or Planned Uses through Fiscal Year 2023

According to USDA, the following components reported no FRT-related activities in fiscal year 2020 or planned uses through fiscal year 2023 listed above:

- Agricultural Marketing Service
- Animal and Plant Health Inspection Service
- Economic Research Service
- Farm Service Agency
- Food and Nutrition Service
- Food Safety and Inspection Service
- Foreign Agricultural Service
- Forest Service
- National Agricultural Statistics Service
- National Institute of Food and Agriculture
- Natural Resources Conservation Service
- Office of the Chief Information Officer
- Office of the Chief Economist
- Office of the General Counsel
- Office of the Inspector General
- Office of Partnerships & Public Engagement
- Office of the Assistant Secretary for Civil Rights
- Office of Tribal Relations
- Office of Homeland Security
- Office of Operations
- Risk Management Agency
- Rural Development
- Rural Utilities Service
- Rural Housing Service
- Rural Business-Cooperative Service

The Department of Commerce reported facial recognition technology (FRT) activities in fiscal year 2020. Specifically, Commerce reported owning one FRT system and conducting FRT-related research and development (R&D) in fiscal year 2020. Commerce also reported plans to use one other FRT system through fiscal year 2023.

Use of FRT Systems in Fiscal Year 2020

Commerce reported it owned one FRT system in fiscal year 2020, for physical security purposes.

Department of Commerce Facial Recognition Technology (FRT) Systems Used in Fiscal Year 2020		
FRT system	**Description of system use**	**Other Commerce and federal users**
Commerce-owned FRT system		
Office of Chief Information Officer TYCO StoneLock infrared biometric facial recognition device	This system uses FRT to control access to a secure data center by infrared facial matching as part of its three-factor authentication process. Commerce reported that it did not continue to use the system after fiscal year 2020.	• None

Source: GAO analysis of survey results. | GAO-21-526

FRT System Obligations in Fiscal Year 2020

Commerce reported it did not obligate funds in fiscal year 2020, because the technology was purchased in 2016.

Research & Development in Fiscal Year 2020

Commerce reported conducting FRT-related R&D, specifically by the National Institute of Standards and Technology (NIST), to support federal agencies and commercial vendors in fiscal year 2020, as follows:

- NIST supported the development of standards and methods in performance measurement, image quality, testing and evaluating technologies, and interoperability for FRT.
- NIST also runs the Facial Recognition Vendor Test program, which evaluates the performance of various facial recognition algorithms against a host of performance metrics, such as accuracy and speed in facial identification/matching, system performance across different demographics, and, more recently, facial detection of individuals wearing masks due to COVID-19.

Commerce reported that while there was no FRT-specific funding, it obligated approximately $500,000 in fiscal year 2020 for biometrics research and standards work at NIST.

Transactions with Nonfederal Entities in Fiscal Year 2020

None reported.

Regulation of Nonfederal Entities' Use of FRT in Fiscal Year 2020

None reported.

Planned Use of FRT Systems through Fiscal Year 2023

Commerce reported plans to use one other FRT system through fiscal year 2023, for physical security purposes.

Department of Commerce Plans for Facial Recognition Technology (FRT) Systems through Fiscal Year 2023

FRT system	Description of system use	Planned new use status
New federal FRT system		
Office of the Chief Information Officer Continuity of Operations Data Center access control system	This system uses infrared facial matching to control access to a secure data center as part of a three-factor authentication process.	Commerce is constructing a data center with this physical security system. Commerce expects to move to the new data center by August 2021.

Source: GAO analysis of survey results. | GAO-21-526

Components with No FRT-Related Activities in Fiscal Year 2020 or Planned Uses through Fiscal Year 2023

According to Commerce, the following components reported no FRT-related activities in fiscal year 2020 or planned uses through fiscal year 2023 listed above:

- Bureau of Economic Analysis
- Bureau of Industry and Security
- U.S. Census Bureau
- Economic Development Administration
- Office of the Under Secretary for Economic Affairs
- International Trade Administration
- Minority Business Development Agency
- National Oceanic and Atmospheric Administration
- National Technical Information Service
- National Telecommunications and Information Administration
- Office of the Inspector General
- Office of the Secretary
- U.S. Patent and Trademark Office

The Department of Defense (DOD) reported facial recognition technology (FRT) activities in fiscal year 2020. Specifically, DOD reported using nine FRT systems and conducting research and development (R&D). DOD also reported plans to use three other FRT systems through fiscal year 2023.

Use of FRT Systems in Fiscal Year 2020

DOD reported it used nine FRT systems in fiscal year 2020, including seven owned by the department and two systems owned by other entities that DOD accessed. Of those nine FRT systems, six are for domestic law enforcement, three are for physical security, five are for national security and defense, and one is for identification card enrollment purposes. Some of these systems are used for multiple purposes.

Department of Defense (DOD) Facial Recognition Technology (FRT) Systems Used in Fiscal Year 2020

FRT system	Description of system use	Other DOD and federal users
DOD-owned FRT systems		
U.S. Army Department of Defense Automated Biometric Identification System (DOD ABIS)	DOD ABIS contains a database of military-collected biometrics of foreign nationals, including faces. It is used to identify threat actors related to terrorism or counterintelligence as well as to research information about a person of interest or identify an individual for an investigative lead.	• DOD (multiple components) • Department of Homeland Security • Department of Justice • Department of State
Defense Manpower Data Center Defense Facial Comparison Tool	The Defense Facial Comparison Tool is used to compare a law enforcement probe photo to a DOD captured photo to identify or verify DOD affiliated individuals.	• DOD (U.S. Navy and U.S. Air Force)
Defense Manpower Data Center Real-Time Automated Personnel Identification System (RAPIDS)	RAPIDS is DOD's enterprise tool for identification card issuance. It uses facial detection to ensure the consistent size of a picture printed on the DOD identification card, during identification card issuance.	• DOD (all components) • Other agencies that issue these identification cards (e.g., the Uniformed Services)
U.S. Navy TacID Guard Dog	TacID Guard Dog is used to monitor camera feeds of individuals seeking access to DOD facilities for possible matches to a watchlist of potential threats.	• None
Pentagon Force Protection Agency M.C. Dean access control device	The device uses FRT to control access to the Pentagon Force Protection Agency door by verifying the identity of authorized individuals.	• DOD (U.S. Army)
U.S. Air Force InCadence Ares Javelin+	Javelin+ is biometrics software installed on mobile devices. The FRT system collects facial images from the devices and submits them to DOD's ABIS to identify an individual in a criminal investigation. This FRT system also performs immediate facial, fingerprint, and iris matching to individuals on watch lists.	• None
U.S. Navy Facial Automated Biometric Identification System (FABIS) Mobile	FABIS Mobile is an FRT phone application that can hold a watchlist. DOD personnel use it during public events, such as airshows, to take photos of individuals who act erratically or suspiciously to determine if they are a potential match to the watchlist. If they are a match, then other biometrics may be taken to verify the individual's identity; if they do not match, DOD personnel send the photo for a manual comparison.	• None
Accessed federal FRT systems		**DOD users**
Department of Justice Next Generation Identification Interstate Photo System (NGI IPS)	NGI IPS can be used by DOD personnel to identify suspected terrorists or for counterintelligence. For example, if an individual is identified as a potential threat actor, DOD may forward the request for a search through NGI IPS.	
Department of Homeland Security Automated Biometric Identification System (IDENT)	IDENT can be used to verify an individual's identity, determine whether an individual in two separate photos is the same person, and compare a probe photo against images stored in IDENT to create a list of potential matches. For example, if an individual is identified by DOD as a potential threat actor, DOD may forward the request for a search through IDENT.	

Source: GAO analysis of survey results. | GAO-21-526

Note: DOD components that reported classified systems not described above, include U.S. Central Command, Defense Intelligence Agency, U.S. European Command, U.S. Indo-Pacific Command, and U.S. Special Operations Command.

FRT System Obligations in Fiscal Year 2020

DOD reported obligating funds to operate its own FRT systems in fiscal year 2020.

- DOD reported obligating funds for four of the seven FRT systems it owned in fiscal year 2020.
  - DOD reported obligating about $3.6 million for DOD ABIS of which a portion is used for facial recognition, about $100,000 for TacID Guard Dog, $865,000 for the Defense Facial Comparison Tool, and $80,000 for FABIS Mobile.
  - DOD did not report obligating funds for Javelin+ or the facial detection portion of RAPIDS, noting that the specific obligation for the FRT portion of that system is unknown because it was included as part of many changes to the RAPIDS system in fiscal year 2020. DOD also reported it obligated no funds for the M.C. Dean access control device, because it was purchased prior to fiscal year 2020.
- DOD reported it did not obligate funds for the two FRT systems it accessed in fiscal year 2020, because federal partners provided them at no cost.

Research & Development in Fiscal Year 2020

DOD reported conducting FRT-related R&D to support its mission needs. For example, DOD reported researching new capabilities for RAPIDS, which would support identity verification during online identification card renewal and personal identification number reset requests.

Transactions with Nonfederal Entities in Fiscal Year 2020

None reported.

Regulation of Nonfederal Entities' Use of FRT in Fiscal Year 2020

None reported.

Planned Use of FRT Systems through Fiscal Year 2023

DOD reported it plans to use three other FRT systems through fiscal year 2023. Specifically, of those three FRT systems, it plans to use two for physical security, one for national security and defense, and one for domestic law enforcement. Of these, one system will be used for multiple purposes.

Department of Defense (DOD) Recognition Technology (FRT) Systems Plans for Facial through Fiscal Year 2023

FRT system	Description of system use	Planned new use status
New federal FRT system		
Defense Advanced Research Projects Agency (DARPA) Visitor Welcome System	DARPA plans to use this FRT to pull a visitor's record automatically, if they have previously been to their Visitor Welcome Center. It will be used to expedite visitor processing, while reducing personnel data entry task errors and duplication.	As of May 2020, DARPA is seeking a license to the algorithm through the U.S. Army.
New access to commercial FRT system		
U.S. Air Force Clearview AI	The U.S. Air Force plans to collect facial images with mobile biometric devices, including phones, to compare against Clearview AI's repository of facial images from open sources for matching individuals.	The U.S. Air Force plans to evaluate the product as part of an operational pilot in June 2020.
Evaluation of federal FRT system		
U.S. Army Automated Installation Entry	The U.S. Army's FRT pilot at Redstone Arsenal performs digitized photo matching in real time at automated access control points using registered Automated Installation Entry system volunteers. It is intended to improve the access vetting process speed and reduce security risks.	The U.S. Army began the pilot in the second quarter of fiscal year 2021.

Source: GAO analysis of survey results. | GAO-21-526

Components with No FRT-Related Activities in Fiscal Year 2020 or Planned Uses through Fiscal Year 2023

According to DOD, the following components reported no FRT-related activities in fiscal year 2020 or planned uses through fiscal year 2023 listed above:

- U.S. Transportation Command
- Department of Defense Chief Information Officer
- Department of Defense Office of the Inspector General

The Department of Energy (DOE) reported facial recognition technology (FRT) activities in fiscal year 2020. Specifically, DOE reported using one FRT system. DOE reported it does not plan to use FRT systems through fiscal year 2023.

Use of FRT Systems in Fiscal Year 2020

DOE reported it owned one FRT system in fiscal year 2020 for physical security purposes.

Department of Energy (DOE) Facial Recognition Technology (FRT) Systems Used in Fiscal Year 2020

FRT system	Description of system use	Other DOE and federal users
DOE-owned FRT system		
National Nuclear Security Administration TacID Guard Dog	TacID Guard Dog performs facial matching and facial detection from live video. DOE uses TacID Guard Dog to monitor entry and exit from controlled locations for personnel accountability and to log individuals' arrivals at an evacuation point.	• None

Source: GAO analysis of survey results. | GAO-21-526

FRT System Obligations in Fiscal Year 2020

DOE reported it purchased TacID Guard Dog in December 2019, and obligated $150,000 for testing the FRT system during fiscal year 2020.

Research & Development in Fiscal Year 2020

None reported.

Transactions with Nonfederal Entities in Fiscal Year 2020

None reported.

Regulation of Nonfederal Entities' Use of FRT in Fiscal Year 2020

None reported.

Planned Use of FRT Systems through Fiscal Year 2023

None reported.

Components with No FRT-Related Activities in Fiscal Year 2020 or Planned Uses through Fiscal Year 2023

According to DOE, the following components reported no FRT-related activities in fiscal year 2020 or planned uses through fiscal year 2023 listed above:

- Office of the Chief Information Officer
- Office of the Inspector General
- Office of Science, including National Laboratories

The Department of Health and Human Services (HHS) reported facial recognition technology (FRT) activities in fiscal year 2020. Specifically, HHS reported using four FRT systems and conducting research and development (R&D). HHS reported plans to expand its use of other FRT systems through fiscal year 2023.

Use of FRT Systems in Fiscal Year 2020

HHS reported it owned three FRT systems and accessed one commercial FRT system in fiscal year 2020. Of those four FRT systems, one was for digital access, one for physical security, and three were for domestic law enforcement purposes. One of these systems was used for multiple purposes.

Department of Health and Human Services (HHS) Facial Recognition Technology (FRT) Systems Used in Fiscal Year 2020

FRT system	Description of system use	Other HHS and federal users
HHS-owned FRT systems		
Centers for Disease Control and Prevention (CDC) AnyVision	AnyVision allows real time facial matching from security camera videos and images. CDC piloted using this system to supplement the manual review and memorization of watchlist faces by security guards at their facilities.	• None
Centers for Medicare and Medicaid Services (CMS) Laptop unlocking	This FRT system allows laptop users that have an appropriate camera to use FRT to unlock their laptop. CMS conducted a small pilot of less than 20 participants using this feature on their laptops.	• None
Office of the Inspector General (OIG) Griffeye Digital Investigate Pro	Griffeye Digital Investigate Pro uses FRT to support automated analysis of digital video and images. The OIG used the system in support of investigations involving child exploitation, child sexual assault, and trafficking to locate or identify victims.	• None
Accessed commercial FRT system		**HHS users**
Clearview AI	Clearview AI is a facial image matching software system that operates as an internet search engine for faces using publicly available images, such as from social media. The OIG conducted an evaluation of the system in an attempt to identify unknown subjects of criminal investigation.	• OIG

Source: GAO analysis of survey results. | GAO-21-526

FRT System Obligations in Fiscal Year 2020

HHS reported obligating funds for two of the three FRT systems it owned in fiscal year 2020.

- HHS reported obligating $1,590 for Griffeye Digital Investigate Pro and $126,896 for AnyVision. HHS reported it obligated no funds for the pilot using facial recognition to unlock laptops because the feature was included with the operating system.

- HHS reported it obligated no funds for its access of Clearview AI, because it was a free trial to evaluate the system.

Research & Development in Fiscal Year 2020

HHS supported research involving facial analysis technology, for purposes other than facial matching. For example, HHS's National Institutes of Health awarded grants for research that use eye tracking as a tool for clinical research, such as characterizing where children with and without autism spectrum disorder looked while following conversations in videos. NIH does not specifically report dollar amounts tied to research eye tracking technology.

Transactions with Nonfederal Entities in Fiscal Year 2020

None reported.

Regulation of Nonfederal Entities' Use of FRT in Fiscal Year 2020

None reported.

Planned Use of FRT Systems through Fiscal Year 2023

HHS reported it plans to use one other FRT system through fiscal year 2023 for domestic law enforcement purposes.

Department of Health and Human Services (HHS) Plans for Facial Recognition Technology (FRT) Systems through Fiscal Year 2023

FRT System	Description of system use	Planned New Use Status
New federal FRT system		
Office of the Inspector General Vintra	Vintra will be used to search surveillance video for investigator-defined actions and objects, such as directional movement, vehicles, or people. The system does not maintain a database of known users but may be able to match a submitted image to an image on the surveillance video.	HHS has tested the system, established security protocols and plans to begin operational usage of Vintra by the end of May 2021.

Source: GAO analysis of survey results. | GAO-21-526

Components with No FRT-Related Activities in Fiscal Year 2020 or Planned Uses through Fiscal Year 2023

According to HHS, the following components reported no FRT-related activities in fiscal year 2020 or planned uses through fiscal year 2023 listed above:

- Administration for Children and Families
- Administration for Community Living
- Agency for Healthcare Research and Quality
- Agency for Toxic Substances and Disease Registry
- Food and Drug Administration
- Health Resources and Services Administration
- Indian Health Service
- Substance Abuse and Mental Health Services Administration

The Department of Homeland Security (DHS) reported facial recognition technology (FRT) activities in fiscal year 2020. Specifically, DHS reported owning four FRT systems and accessing three federal FRT systems, FRT systems in 17 states and five localities, and two commercial FRT systems. DHS also reported conducting research and development (R&D) related to FRT, entering into transactions with nonfederal entities for FRT, and regulating nonfederal entities use of FRT. DHS also reported it plans to use five other FRT systems through fiscal year 2023.

Use of FRT Systems in Fiscal Year 2020

DHS owned four federal FRT systems and accessed multiple systems owned by other entities. DHS used four FRT systems for domestic law enforcement, six for border and transportation security, and four for national security and defense purposes. Some of these systems were used for multiple purposes. DHS had access to at least 24 state, local, and commercial FRT systems for domestic law enforcement, for border and transportation security, and for national security and defense purposes.

Department of Homeland Security (DHS) Facial Recognition Technology (FRT) Systems Used in Fiscal Year 2020

FRT system	Description of system use	Other DHS and federal users
DHS-owned FRT systems		
U.S. Customs and Border Protection Automated Targeting System (ATS)	U.S. Customs and Border Protection uses ATS's FRT for the following populations: (1) individuals seeking to enter or exit the United States whose names appear on a flight or vessel manifests, or voluntary manifests submitted by bus or rail manifest ("manifested travelers"); (2) individuals applying for CBP programs facilitating travel to the United States, and (3) subjects of interest who require additional research and analysis. It matches photos for these three populations against a predetermined gallery of photos associated with derogatory information.	• Department of Justice • Department of State
U.S. Customs and Border Protection Traveler Verification Service (TVS)	TVS uses facial recognition to verify traveler identities upon arrival at or departure from ports of entry. TVS compares a live photo of a traveler against a gallery of photos (e.g., passport photos) in DHS databases.	• DHS (Transportation Security Administration)
Office of Biometric Identity Management Automated Biometric Identification System (IDENT)	IDENT offers facial recognition services to partners to verify an individual s identity, determine whether an individual in two separate photos is the same, and compare a probe photo against images stored in IDENT for potential matches.	• DHS (all components) • Department of Justice • Department of Defense
Transportation Security Administration Self-Service Version of Credential Authentication Technology with Camera (CAT-2) and AutoCat[a]	CAT-2 machines provide facial matching services to assist with identity verification of travelers by capturing live images at airport checkpoints. AutoCat is an electronic gate version of CAT-2.	• DHS (Science and Technology Directorate)

Accessed federal FRT systems		DHS users
Department of Defense Automated Biometric Identification System (DOD ABIS)	DHS uses DOD ABIS's facial recognition to identify foreign nationals connected to a national security investigation or to a known terrorist organization.	• U.S. Immigration and Customs Enforcement
Department of Justice Next Generation Identification Interstate Photo System (NGI IPS)	DHS uses NGI IPS to identify individuals of interest. Specifically, U.S. Customs and Border Protection submits probe photos via ATS to IDENT and are forwarded to NGI IPS. NGI IPS returns a list of potential matching photos.	• Office of Biometric Identity Management • U.S. Customs and Border Protection
Department of State Integrated Biometric System (IBS)	DHS uses IBS to identify visa applicants for travel documents and individuals involved in identity theft and benefit fraud investigations. DHS employees have direct and indirect access to IBS—performing facial recognition searches in IBS while others submit photos for matching from apprehensions and bookings, among others.	• U.S. Citizenship and Immigration Services • U.S. Customs and Border Protection • U.S. Immigration and Customs Enforcement
Accessed state and local FRT systems		**DHS users**
15 states: Alabama, Arizona, Delaware, Indiana, Kansas, Kentucky, Michigan, Mississippi, Missouri, Nebraska, North Dakota, South Dakota, Utah, Virginia, and West Virginia 4 localities: Chicago, Greater Cincinnati, Delaware Valley, Southeast Florida	DHS's Homeland Security Information Network (HSIN) is a system for trusted sharing of Sensitive But Unclassified information between federal, state, local, territorial, tribal, international, and private sector partners. HSIN contains a mechanism to request third party facial recognition searches through the listed state and local entities, such as fusion centers.	• U.S. Customs and Border Protection • U.S. Immigration and Customs Enforcement
Michigan Law Enforcement Information Network (MLEIN)	DHS uses MLEIN to identify individuals involved in a crime. DHS submits U.S. Border Patrol apprehension photos and others to the Statewide Network of Agency Photos.	• U.S. Customs and Border Protection
New York State Intelligence Center Photo Imaging Mugshot System (PIMS)	DHS uses PIMS to identify unknown individuals of interest in state and federal cases, including individuals who have been deported and re-entered the country. DHS has direct access to PIMS through Border Patrol Agents working at the New York State Intelligence Center.	• U.S. Customs and Border Protection
Ohio Law Enforcement Gateway (OHLEG)	DHS uses OHLEG to identify individuals involved in a crime. DHS submits photos, such as U.S. Border Patrol apprehension photos.	• U.S. Customs and Border Protection
Pinellas County Face Analysis Comparison and Examination System (FACES)	DHS uses FACES to identify unknown individuals to support operations, criminal investigations, and administrative cases. DHS submits photos to the Pinellas County Sheriff's Office.	• U.S. Customs and Border Protection
Accessed commercial FRT systems		**DHS users**
Clearview AI	DHS uses Clearview AI to identify unknown individuals of interest in state, federal, and international cases, such as child exploitation cases. DHS may submit photos (e.g., surveillance photos) or receive requests to match photos against the Clearview AI database.	• U.S. Customs and Border Protection • U.S. Immigration and Customs Enforcement • U.S. Secret Service
Vigilant Solutions	DHS uses Vigilant Solutions to identify individuals involved in a crime and submits photos for matching.	• U.S. Customs and Border Protection

Source: GAO analysis of survey results. | GAO-21-[illegible]

Note: DHS also reported using classified systems.

[a]This system was only tested and not deployed.

FRT System Obligations in Fiscal Year 2020

DHS reported obligating funds to operate its own FRT systems or access other ones, including state and commercial systems, in fiscal year 2020.

- DHS reported obligating funds for three of the four FRT systems it owns in fiscal year 2020. Specifically, DHS reported obligating $7.96 million for the annual maintenance cost of the license for IDENT, $61.5 million for TVS, and $2.5 million for CAT-2. DHS reported it obligated no funds for ATS, because ATS is funded through the license for IDENT.
- DHS reported obligating funds for two FRT systems it accessed in fiscal year 2020. Specifically, DHS reported obligating about $23.67 million for HSIN and $31,592 to fund approximately seven licenses under its contract for access to the Michigan State Police's MLEIN. For the remaining federal, state, and local FRT systems it accessed,
    - DHS reported obligating no funds to access three federal FRT systems—DOD ABIS, NGI IPS, and IBS—because those systems are funded by DOD, DOJ, and State, respectively.
    - DHS reported obligating no funds for state and local FRT systems, because they were accessed at no cost through respective state and local owners.
- DHS reported obligating funds to access one of two commercial FRT systems in fiscal year 2020. Specifically, DHS' Immigration and Customs Enforcement reported obligating $214,000 to purchase Clearview AI licenses, while Secret Service has not yet made a decision to purchase services after conducting a pilot in April 2019. U.S. Customs and Border Protection accessed Clearview AI at no cost through an agent stationed at the New York State Intelligence Center. Finally, DHS reported it has limited access to Vigilant Solutions at no cost.

Research & Development in Fiscal Year 2020

DHS reported conducting R&D for testing on one FRT system and additional research to improve face detection, analysis, and matching capabilities:

- The Transportation Security Administration reported obligating over $2.17 million for independent testing and planning for CAT-2. The tests will validate facial matching performance through a series of pilots of the Transportation Security Administration's specific use case at checkpoints.
- The Science and Technology Directorate reported obligating $1.2 million to sponsor Biometric Technology Rallies, which are ongoing industry events with challenges to develop innovative solutions for biometric collection and matching, including facial recognition, and the reliability of collecting and matching information on travelers wearing masks during COVID-19.
- The Science and Technology Directorate and Office of Biometric Identity Management reported obligating about $553,000 for National Science Foundation's Center for Identification Technology Research university grants. Specifically, S&T sponsors R&D and awards grants to university partners that focus on work related to face detection, analysis, and matching. OBIM sponsors R&D with a portion of work including fingerprint matching, presentation attack detection, and matching of biometrics from juveniles.
- The Science and Technology Directorate reported obligating $300,000 to sponsor R&D on face detection, analysis, and matching through its interagency agreement with the National Institute of Standards and Technology.

Transactions with Nonfederal Entities in Fiscal Year 2020

DHS reported that it entered into agreements with foreign governments and private entities, and awarded grants to local governments and other entities:

- The Transportation Security Administration reported it had a Cooperative Research and Development Agreement with a technology company and an air carrier focused on ways to make mobile digital identity credentials (e.g., mobile digital driver's licenses) interoperable with Transportation Security Administration's checkpoint security systems.

    - The Science and Technology Directorate entered into project arrangements with two foreign governments—Australia and the United Kingdom—related to the assessment of face recognition software.
    - DHS awarded a contract to the Lehigh County, Pennsylvania District Attorney's Office to enhance the U.S. Immigration and Customs Enforcement Homeland Security Investigation's future access to the Gang Intelligence Application. DHS reported spending about $200,000 on the application in fiscal year 2020.
    - The Federal Emergency Management Agency reported it obligated about $1.25 million across several grants that included facial recognition, such as equipment for emergency responders and state and local law enforcement.

Regulation of Nonfederal Entities' Use of FRT in Fiscal Year 2020

DHS reported regulated a nonfederal entity's use of FRT. Specifically, the Transportation Security Administration issued security program amendments to Delta Air Lines to permit the use of facial identification technology to identify passengers checking baggage for air transportation.

Planned Use of FRT Systems through Fiscal Year 2023

DHS reported it plans to use five other FRT systems through fiscal year 2023. Of those five FRT systems, three will be used for border and transportation security, two will be used for national security and defense, and two will be used for domestic law enforcement purposes. Some of these systems will be used for multiple purposes.

Department of Homeland Security (DHS) Plans for Facial Recognition Technology (FRT) Systems through Fiscal Year 2023

FRT system	Description of system use	Planned new use status
New access to local FRT system		
Lehigh County, Pennsylvania District Attorney's Office Gang Intelligence Application	U.S. Immigration and Customs Enforcement plans to develop facial recognition access to its data in an automated criminal justice information system.	U.S. Immigration and Customs Enforcement is working with the Lehigh County District Attorney's Office to complete access by May 2021.
Evaluation of federal FRT system		
U.S. Customs and Border Protection Traveler Verification Service (TVS)	The Transportation Security Administration plans to initiate a new pilot of U.S. Customs and Border Protection's TVS to evaluate use of biometric technology, including facial recognition, to automate identity verification at checkpoints and modernize screening.	The Transportation Security Administration began a pilot at the Detroit Metropolitan Wayne County Airport in March 2021 with U.S. Customs and Border Protection.

Department of Homeland Security (DHS) Plans for Facial Recognition Technology (FRT) Systems through Fiscal Year 2023

FRT system	Description of system use	Planned new use status
Transportation Security Administration Credential Authentication Technology with Camera System (CAT-2) and AutoCAT	The Transportation Security Administration plans to test this technology to automate identity verification at checkpoints and modernize screening of travelers.	The Transportation Security Administration began demonstrating CAT-2 at Ronald Regan Washington National Airport in August 2020. In March 2021, it started field site testing of CAT-2 at additional airport checkpoints to identify, evaluate, and mitigate system performance issues across diverse operational environments and passenger demographics. Data collected during field tests will be used for qualitative and quantitative analysis by the Science and Technology Directorate. It is also testing and monitoring AutoCAT at the TSA Systems Integration Facility in preparation for future field pilots.
Upgrade of federal FRT system		
Office of Biometric Identity Management Homeland Advanced Recognition Technology (HART)	HART will replace IDENT's capabilities initially, to include FRT, and additional capabilities will be added in subsequent years.	The Office of Biometric Identity Management plans to replace IDENT with HART by December 2021.

Source: GAO analysis of survey results. | GAO-21-526

Note: We included DHS's plans to pilot an additional federal FRT system, the details of which are sensitive, in the reported number of planned use of FRT systems.

Components with No FRT-Related Activities in Fiscal Year 2020 or Planned Uses through Fiscal Year 2023

According to DHS, the following components reported no FRT-related activities in fiscal year 2020 or planned uses through fiscal year 2023 listed above:

- U.S. Coast Guard
- Cybersecurity and Infrastructure Security Agency
- Federal Law Enforcement Training Center
- Countering Weapons of Mass Destruction
- Office of the Inspector General
- Office of Operations Coordination

The Department of the Interior reported facial recognition technology (FRT) activities in fiscal year 2020. Specifically, Interior reported accessing two FRT systems in fiscal year 2020 and planning to use two FRT systems through fiscal year 2023.

Use of FRT Systems in Fiscal Year 2020

Interior reported it accessed one locally owned FRT system and one commercial FRT system in fiscal year 2020, for domestic law enforcement.

Department of the Interior Facial Recognition Technology (FRT) Systems Used in Fiscal Year 2020

FRT system	Description of system use	Other Interior and federal users
Accessed Local FRT system		**Interior users**
National Capital Region Facial Recognition Investigative Leads System (NCRFRILS)	NCRFRILS is an FRT system that contains copies of information, including photos, from participating law enforcement agencies. Interior requested a third party search of NCRFRILS to compare an image obtained from twitter (as an example) against the photo database to generate investigate leads.	• Interior (U.S. Park Police)
Accessed commercial FRT system		**Interior users**
Clearview AI	Clearview AI is an FRT system that can identify an individual through facial matching by comparing a photo against its facial image database. Interior uses Clearview AI to verify the identity of an individual involved in a crime and research information on a person of interest. Interior may submit photos (e.g., surveillance photos) for matching against the Clearview AI's repository of facial images from open sources. U.S. Park Police reported it stopped using Clearview AI as of June 2020.	• Interior (U.S. Park Police)

Source: GAO analysis of survey results. | GAO-21-526

FRT System Obligations in Fiscal Year 2020

Interior reported it obligated no funding to access NCRFRILS in fiscal year 2020, because Interior (specifically, the U.S. Park Police) accessed this system through a third party—the Maryland National Capital Park Police. In addition, Interior reported it obligated no funding to access Clearview AI in fiscal year 2020, because it was a free trial of Clearview AI.

Research & Development in Fiscal Year 2020

None reported.

Transactions with Nonfederal Entities in Fiscal Year 2020

None reported.

Regulation of Nonfederal Entities' Use of FRT in Fiscal Year 2020

None reported.

Planned Use of FRT Systems through Fiscal Year 2023

Interior reported it plans new access to a local FRT system and a commercial FRT system through fiscal year 2023, for domestic law enforcement purposes.

Department of the Interior Plans for Facial Recognition Technology (FRT) Systems through Fiscal Year 2023		
FRT system	**Description of system use**	**Planned new use status**
New access to local FRT system		
National Capital Region Facial Recognition Investigative Leads System (NCRFRILS)	NCRFRILS is an FRT system that contains copies of information, including photos, from participating law enforcement agencies. Interior requested a third party search of NCRFRILS to compare an image obtained from twitter (as an example) against the photo database to generate investigate leads.	In April 2021, U.S. Park Police officials told us their plans to establish a memorandum of understanding for direct access to NCRFRILS.
New access to commercial FRT system		
Clearview AI	Clearview AI is an FRT system that can identify an individual through facial matching by comparing a photo against its facial image database. Interior will use Clearview AI to verify the identity of an individual involved in a crime and research information on a person of interest. Interior may submit photos (e.g., surveillance photos) for matching against the Clearview AI repository of facial images from open sources	Interior reported its U.S. Fish and Wildlife Service began using a trial version of Clearview AI in May 2020, and purchased an annual subscription in June 2020.

Source: GAO analysis of survey results. | GAO-21-526

Components with No FRT-Related Activities in Fiscal Year 2020 or Planned Uses through Fiscal Year 2023

According to Interior, the following components reported no FRT-related activities in fiscal year 2020 or planned uses through fiscal year 2023 listed above:

- Bureau of Indian Affairs
- Bureau of Indian Education
- Bureau of Land Management
- Bureau of Ocean Energy Management
- Bureau of Reclamation
- Bureau of Safety and Environmental Enforcement
- Bureau of Trust Funds Administration
- Office of the Assistant Secretary for Policy, Management and Budget
- Office of the Inspector General
- Office of the Secretary
- Office of Surface Mining Reclamation and Enforcement
- Office of the Solicitor
- U.S. Geological Survey

The Department of Justice (DOJ) reported facial recognition technology (FRT) activities in fiscal year 2020. Specifically, DOJ reported using federal, state, local, and commercial FRT systems, conducting research and development (R&D) related to FRT, and entering into transactions with nonfederal entities for FRT. DOJ also reported plans to expand its use of FRT through fiscal year 2023.

Use of FRT Systems in Fiscal Year 2020

DOJ reported it used 11 federal and commercial FRT systems in fiscal year 2020, in addition to a number of state and local systems. Of those 11 FRT systems, DOJ used eight for domestic law enforcement, one for physical security, two for national security and defense, one for video management, and two for educational purposes. Some of these systems were used for multiple purposes. All of the state and local systems were used for domestic law enforcement and some national security and defense purposes.

Department of Justice (DOJ) Facial Recognition Technology (FRT) Systems Used in Fiscal Year 2020		
FRT system	**Description of system use**	**Other DOJ and federal users**
DOJ-owned FRT systems		
Federal Bureau of Investigation Horus	Horus is being tested to determine if it can be used to improve the accuracy and efficiency of one-to-one comparison processes by serving as an aid to examiners. It is also used in educational settings to demonstrate how FRT works.	• None
Federal Bureau of Investigation Next Generation Identification Interstate Photo System (NGI IPS)	NGI IPS uses facial matching from an unknown image of interest to law enforcement against mugshots in the photo database to generate possible matches for investigators.	• Department of Homeland Security • Department of Defense
Federal Bureau of Investigation RankOne	Similar to Horus, this system is being tested to determine if it can be used to improve the accuracy and efficiency of one-to-one comparison processes by serving as an aid to examiners. It is also used in educational settings to demonstrate how FRT works.	• None
Federal Bureau of Prisons Facial Recognition Access Control System	This system uses FRT to control access by identifying and verifying individuals entering onsite secure network operations centers at federal prisons. Federal Bureau of Prisons officials noted the system will not be in use after fiscal year 2022.	• None
U.S. National Central Bureau (USNCB) International Criminal Police Organization (INTERPOL) Facial Recognition System (IFRS)	At the request of U.S. law enforcement officials, including state, local, tribal and federal authorities, the U.S. National Central Bureau can send photos, such as missing persons and suspects, in investigative cases for comparison against the holdings in IFRS. IFRS contains facial images received from more than 160 countries, and member countries can provide facial images for matching. These results are returned to the requesting country and also to the country that provided the images.	• Any U.S. law enforcement agency, including federal, via the U.S. National Central Bureau
U.S. Marshals Service Axon Facial Detection	Axon Facial Detection is used to review footage from body-worn cameras for faces (i.e., face detection) that a human then selects/deselects for redaction across multiple video frames.	• None
Accessed federal FRT systems		**DOJ users**
Department of Defense Automated Biometric Identification System (DOD ABIS)	DOD ABIS is owned and operated by the Department of Defense and is used to identify foreign nationals connected to a national security investigation, such as a known terrorist organization, or suspected of criminal activity. DOJ can request an image search on DOD ABIS through FBI's NGI IPS.	• Federal Bureau of Investigation
Department of State Integrated Biometric System (IBS)	IBS is owned and operated by the Department of State and is used to identify visa applicants for travel documents and individuals involved in identity theft and benefit fraud investigations. Specifically, FBI has direct access to perform facial recognition searches within the Department of State Consular Consolidated Database's visa holdings, and FBI's Facial Analysis, Comparison, and Evaluation (FACE) Services has indirect access to passport photos. FACE Services must request a Department of State passport officer perform facial recognition searches of passport photos on the FBI's behalf.	• Federal Bureau of Investigation

Accessed state and local FRT systems		DOJ users
21 states: Alabama, Arizona, Arkansas, Colorado, Delaware, Idaho, Illinois, Indiana, Iowa, Kentucky, Maryland, Michigan, Nebraska, New Mexico, North Carolina, North Dakota, Pennsylvania, South Carolina, Tennessee, Texas, and Utah	The FBI's FACE Services is an internal provider of facial recognition searches and requests for FBI investigations. FACE Services examiners have direct or third party access through memorandums of understandings with 21 different states' FRT systems. FACE Services also has direct and indirect (i.e., by request) access to various holdings within the Department of State's IBS and indirect access to DOD ABIS.	• Federal Bureau of Investigation
15 states: Alabama, Arizona, Delaware, Indiana, Kansas, Kentucky, Michigan, Mississippi, Missouri, Nebraska, North Dakota, South Dakota, Utah, Virginia, and West Virginia 4 localities: Chicago, Greater Cincinnati, Delaware Valley, Southeast Florida	DHS's Homeland Security Information Network (HSIN) is a system for trusted sharing of Sensitive But Unclassified information between federal, state, local, territorial, tribal, international and private sector partners. HSIN contains a mechanism to request third party facial recognition searches through the listed state and local entities, such as fusion centers.	• U.S. Marshals Service
Other states: Arkansas, Arizona, Ohio, Pennsylvania, and Tennessee Other localities: Pinellas County, Florida and San Diego, California	DOJ requested third-party searches of these systems separately by the state or local owner in order to identify fugitives, individuals involved in a crime, or people using fraudulent identification (i.e., aliases), among other persons of interest.	• U.S. Marshals Service
National Capital Region Facial Recognition Investigative Leads System (NCRFRILS)	Authorized DOJ employees can directly search NCRFRILS for matches against photos from participating local law enforcement agencies. For example, an examiner reviewed video evidence from a retail store's surveillance cameras with facial recognition software, comparing the still image from the video with regional booking photos.	• Bureau of Alcohol, Tobacco, Firearms, and Explosives • U.S. Marshals Service

Research & Development in Fiscal Year 2020

DOJ reported conducting and supporting R&D to support its mission needs. Specifically, the FBI reported obligating $1.56 million for several types of FRT research, as follows:

- An interagency agreement with the National Institute of Standards and Technology for (1) testing and evaluation of current industry face image quality tools and identification of best practices; (2) a facial recognition algorithm benchmark for testing overall accuracy and continued analysis of the effect demographics has on accuracy; and (3) benchmark testing technology to detect face image manipulation, such as a deepfake, with the goal of discerning industry capabilities.
- Applied research at the West Virginia University Research Corporation into (1) the relationship between skin tone and false match rates in facial recognition algorithms and (2) assessing the capabilities and limitations of current synthetic face detection, such as deepfakes, and developing synthetic face detection software prototypes.

- FBI's Operational Technology Division is conducting applied research on facial matching. For example, the Horus and RankOne systems were tested for the potential benefits of combining FRT systems with trained forensic examiners for verification.

Transactions with Nonfederal Entities in Fiscal Year 2020

DOJ reported that it awarded an $836,000 grant to the Police Foundation for the development of techniques to automate analysis of body worn camera audio and video data of police and community interactions. In particular, these techniques could (1) allow an evaluation of officers' adherence to principles of procedural justice and (2) validate the ratings generated by the automated process using a randomized control trial comparing software ratings of videos to evaluations performed by human raters under conditions of high and low procedural justice.

Regulation of Nonfederal Entities' Use of FRT in Fiscal Year 2020

None reported.

Planned Use of FRT Systems through Fiscal Year 2023

DOJ reported it plans to use two other FRT systems through fiscal year 2023. Of these, one will be used for border and transportation security, and one will be used for physical security purposes.

Department of Justice (DOJ) Plans for Facial Recognition Technology (FRT) Systems through Fiscal Year 2023

FRT system	Description of system use	Planned new use status
New Federal FRT systems		
Drug Enforcement Administration (DEA) Security video management system	DEA plans to use the facial recognition component of a security video management system to control access, monitor, and surveil its headquarters facility.	This system is not operational as of June 2021 because DEA's headquarters facility is being renovated. DEA plans to begin using the FRT features after completing a privacy assessment.
U.S. Marshals Service (USMS) Justice Prisoner and Alien Transportation System (JPATS) Mobile Application	USMS plans to develop software to perform touchless prisoner identity verification, such as during prisoner transport, searching for a match against booking and prisoner photos within the JPATS mobile application for transportation security purposes.	USMS planned to begin using the JPATS Mobile Application in mid-to-late 2021, but has been delayed because the bid was protested.

Source: GAO analysis of survey results. | GAO-21-526

- U.S. Parole Commission
- Executive Office for Immigration Review
- Executive Office for Organized Crime Drug Enforcement Task Forces
- Executive Office for U.S. Trustees
- Professional Responsibility Advisory Office

The Department of State reported facial recognition technology (FRT) activities in fiscal year 2020. Specifically, State reported owning one FRT system and accessing one FRT system in fiscal year 2020. State also reported conducting FRT-related research and development (R&D), and entering into transactions with nonfederal entities for FRT. State reported it plans to use one other FRT system through fiscal year 2023.

Use of FRT Systems in Fiscal Year 2020

State reported it owned one FRT system and accessed one other FRT system in fiscal year 2020, for border and transportation security and national security and defense purposes.

Department of State Facial Recognition Technology (FRT) Systems Used in Fiscal Year 2020

FRT system	Description of system use	Other State and federal users
State-owned FRT system		
Bureau of Consular Affairs Integrated Biometric System (IBS)	State uses IBS to verify an applicant's identity or determine whether an individual has applied for a visa. State uses FRT to search visa and passport photos contained in IBS for matches with individuals who previously applied for travel documents or may be involved in visa fraud. Potential matches from IBS are sent to the Kentucky Consular Center and National Visa Center for review before a final determination is made by consular officers at posts or passport specialists at passport agencies and centers.	• Department of State (Bureau of Diplomatic Security) • Department of Justice • Department of Homeland Security
Accessed federal FRT system		**State users**
Department of Defense Automated Biometric Identification System (DOD ABIS)	State uses DOD ABIS to verify the authenticity of travel documents. State submits multimodal biometric files (e.g., fingerprints and face and iris scans) to DOD ABIS through a U.S. Special Operations Command portal or the Department of Justice s Next Generation Identification Interstate Photo System.	• Bureau of Diplomatic Security

Source: GAO analysis of survey results. | GAO-21-526

FRT System Obligations in Fiscal Year 2020

In fiscal year 2020, State reported it obligated over $8.5 million for IBS. In addition, State reported it obligated no funds to access DOD ABIS, because DOD funded the FRT system.

Research & Development in Fiscal Year 2020

State reported conducting the following FRT-related R&D to enhance facial analysis and facial matching in fiscal year 2020, as follows:

- State's Bureau of Consular Affairs reported conducting R&D to enhance facial analysis, including morphing detection (e.g., making a synthetic face image using two faces to allow both individuals to use the image for identification purposes) and the impact of aging on the accuracy of facial recognition algorithms, such as for children's passports. State reported this R&D will contribute to International Organization for Standardization and International Civil Aviation Organization image standards for travel documents. State reported obligating over $1.7 million for these projects in fiscal year 2020.
- State's Bureau of Counterterrorism reported conducting R&D to enhance the facial matching of screening technology used in the Personal Identification Secure Comparison and Evaluation System (PISCES) border management system. The system matches images of individuals against passport images and a repository of images of suspicious individuals. State reported obligating approximately $3 million for a larger effort to improve program technology, including the enhanced screening technology.

Transactions with Nonfederal Entities in Fiscal Year 2020

State reported it entered into the following transactions to assist foreign governments with purchasing and using previously donated FRT equipment:

- State's Bureau of International Narcotics and Law Enforcement reported entering into an interagency agreement and contract with Mexico's National Migration Institute. Specifically, State assisted with building the institute's capacity to collect, store, and share biometric data on third-country nationals with donated FRT equipment. In addition, State reported it funded technical advisors to train Mexican government employees to use the equipment. State reported obligating approximately $2.6 million for the technical assistance and training.
- State's Bureau of International Narcotics and Law Enforcement also reported it obligated approximately $333,000 to purchase 10 workstations for the Guatemalan Immigration Institute. The institute plans to install the workstations at three locations—the La Aurora International Airport, the Guatemalan Immigration Institute Detention Facility, and the Valle Nuevo port of entry on the El Salvador-Guatemala border.

Regulation of Nonfederal Entities' Use of FRT in Fiscal Year 2020

None reported.

Planned Use of FRT Systems through Fiscal Year 2023

State reported it plans to use one other FRT system through fiscal year 2023 for border and transportation security.

Department of State Plans for Facial Recognition Technology (FRT) Systems through Fiscal Year 2023

FRT system	Description of system use	Planned new use status
New federal FRT system		
Bureau of Counterterrorism Personal Identification Secure Comparison and Evaluation System (PISCES)	The FRT system matches individuals against passport images and a repository of known or suspected terrorists using the system. State reported plans to incorporate enhanced screening technologies in PISCES to ensure foreign partners under the Terrorist Interdiction Program are able to protect themselves from attempts by terrorists to enter, transit, or depart their country.	State plans to pilot the new FRT system in the summer of 2021. If the testing is successful, State will deploy the software to operational locations.

Source: GAO analysis of survey results. | GAO-21-526

Components with No FRT-Related Activities in Fiscal Year 2020 or Planned Uses through Fiscal Year 2023

According to State, the following components reported no FRT-related activities in fiscal year 2020 or planned uses through fiscal year 2023 listed above:

- Office of the Secretary of State
- Office of the Deputy Secretary
- Office of the Under Secretary for Arms Control and International Security
- Office of the Under Secretary for Economic Growth, Energy, and Environment
- Office of the Under Secretary for Political Affairs
- Office of the Under Secretary for Public Diplomacy and Public Affairs

The Department of Transportation (DOT) reported conducting facial recognition technology (FRT) related research and development (R&D) activities in fiscal year 2020. Specifically, DOT used FRT to conduct human factors research in a variety of transportation-related areas. DOT reported it did not have any other FRT activities in fiscal year 2020, and does not plan to use FRT systems through fiscal year 2023.

Use of FRT Systems in Fiscal Year 2020

None reported.

FRT System Obligations in Fiscal Year 2020

None reported.

Research & Development in Fiscal Year 2020

DOT reported the following R&D projects in fiscal year 2020 that used FRT, specifically facial detection and analysis, to conduct human factors research, such as:

- The Federal Motor Carrier Safety Administration conducted human factors research involving commercial motor vehicle drivers' behavior and safety performance. For example, an onboard monitoring system analyzes a driver's face to identify fatigue and distraction while driving. The FRT has eye glance analysis, which measures visual attention or inattention, rates drowsiness, and measures the percent of eye closure as an indicator of fatigue.
- The Federal Highway Administration used FRT at its lab, which has a simulator and test vehicle, to conduct human factors research. They use the results to incorporate highway driver needs into roadway design, construction, repair, and improvement. In addition, it obligated $150,580 through an interagency agreement to the National Science Foundation's Big Data Hubs program to support research into better ways to protect the privacy of drivers who participated in a mandated naturalistic driver study that recorded their in-cabin behavior. This research is intended to improve other researchers' access to the driver data, helping them use it to improve traffic safety. It also obligated $300,000 to Oak Ridge National Laboratory, which included facial detection research.
- The Federal Railroad Administration reported operating the Cab Technology Integration Lab—a full-sized locomotive cab simulator—that uses an eye tracking device and software to determine what a train engineer is looking at on a display screen or when scanning the environment outside the cab. This data helps researchers determine what objects in the environment capture the engineer's attention while driving a train.
- The Federal Aviation Administration reported conducting human factors research that uses eye tracking technology to observe how air traffic controllers scan their instruments to determine eye motion workload and the time spent looking at each object. This research is intended to ensure that systems that include human operators and maintainers perform as effectively and safely as possible.

DOT reported it could not provide specific obligations for individual R&D projects that used FRT, because the programs do not track obligations at that level.

Transactions with Nonfederal Entities in Fiscal Year 2020

None reported.

Regulation of Nonfederal Entities' Use of FRT in Fiscal Year 2020

None reported.

Planned Use of FRT Systems through Fiscal Year 2023

None reported.

Components with No FRT-Related Activities in Fiscal Year 2020 or Planned Uses through Fiscal Year 2023

According to DOT, the following components reported no FRT-related activities in fiscal year 2020 or planned uses through fiscal year 2023 listed above:

- Office of the Secretary of Transportation
- Office of Inspector General
- National Highway Traffic Safety Administration
- Federal Transit Administration
- Pipeline and Hazardous Materials Safety Administration
- Great Lakes Saint Lawrence Seaway Development Corporation
- Maritime Administration

The Department of the Treasury reported facial recognition technology (FRT) activities in fiscal year 2020. Specifically, Treasury reported accessing two FRT systems. Treasury also reported plans to use three other FRT systems through fiscal year 2023.

Use of FRT Systems in Fiscal Year 2020

Treasury reported it accessed two FRT systems in fiscal year 2020, one for digital access and one for domestic law enforcement purposes.

Department of the Treasury Facial Recognition Technology (FRT) Systems Used in Fiscal Year 2020

FRT system	Description of system use	Other Treasury and federal users
Accessed federal FRT system		**Treasury users**
General Services Administration login.gov	This login.gov pilot used FRT that compared two photos to verify the identity of an individual accessing the website or application. Login.gov takes a picture of the individual and their photo identification to determine a match using FRT.	• Internal Revenue Service (IRS)
Accessed commercial FRT system		**Treasury users**
Vendor facial recognition search services	A third-party vendor performed facial recognition searches on behalf of the IRS for domestic law enforcement purposes. Additional details on the search are sensitive.	• IRS

Source: GAO analysis of survey results. | GAO-21-526

FRT System Obligations in Fiscal Year 2020

Treasury reported it obligated no funding to access the two FRT systems in fiscal year 2020. Specifically, GSA funded its login.gov FRT system, and a vendor provided a demonstration of their FRT system at no additional cost.

Research & Development in Fiscal Year 2020

None reported.

Transactions with Nonfederal Entities in Fiscal Year 2020

None reported.

Regulation of Nonfederal Entities' Use of FRT in Fiscal Year 2020

None reported.

Planned Use of FRT Systems through Fiscal Year 2023

Treasury reported it plans to use three FRT systems through fiscal year 2023 for domestic law enforcement purposes.

Department of the Treasury Plans for Facial Recognition Technology (FRT) Systems through Fiscal Year 2023

FRT system	Description of system use	Planned new use status
New federal FRT system		
U.S. Treasury Inspector General for Tax Administration (TIGTA) Facebox	Facebox identifies facial images of similar persons linked to multiple investigations through an online storage locker that contains information, including photos from other investigations. The FRT system notifies the investigators of potential matches.	TIGTA purchased Facebox in June 2020, and plans to begin use of the FRT system in December 2020.

New access to federal FRT system		
Internal Revenue Service (IRS) Memoranda of Understanding (MOUs) to access federal FRT systems	The MOUs will allow IRS agents direct access to other federal biometric databases, including facial recognition. The additional access will assist IRS agents with criminal investigations. For example, an IRS agent may submit a surveillance photo to the National Forensic Laboratory, which will conduct a facial recognition search against one of the FRT systems.	The IRS's National Forensic Laboratory is in the process of establishing MOUs with the Departments of Defense, Homeland Security, and Justice. As of April 2021, the IRS was determining whether separate MOUs are needed.
Evaluation of commercial FRT system		
IRS Vendor facial recognition search services	A third-party vendor performed facial recognition searches on behalf of the IRS for domestic law enforcement purposes. Additional details on the search are sensitive.	IRS plans to conduct an additional pilot pending additional funding.

Source: GAO analysis of survey results. | GAO-21-526

Components with No FRT-Related Activities in Fiscal Year 2020 or Planned Uses through Fiscal Year 2023

According to Treasury, the following components reported no FRT-related activities in fiscal year 2020 or planned uses through fiscal year 2023 listed above:

- Alcohol and Tobacco Tax and Trade Bureau
- Bureau of Engraving and Printing
- Bureau of the Fiscal Service
- Financial Crimes Enforcement Network
- Office of Inspector General
- Office of the Comptroller of the Currency
- U.S. Mint
- Treasury Departmental Offices

The Department of Veterans Affairs (VA) reported facial recognition technology (FRT) activities in fiscal year 2020. Specifically, VA reported conducting FRT-related research and development (R&D), and entering into transactions with nonfederal entities for FRT. VA reported plans to use two other FRT systems through fiscal year 2023.

Use of FRT Systems in Fiscal Year 2020

None reported.

FRT System Obligations in Fiscal Year 2020

None reported.

Research & Development in Fiscal Year 2020

VA reported using FRT as a tool to conduct research for purposes other than facial matching. Specifically, VA reported using eye tracking to conduct clinical research for treating post-traumatic stress disorder. The eye tracking system evaluates pupil response to evaluate impairment. VA reported obligating $22,840 for the equipment.

Transactions with Nonfederal Entities in Fiscal Year 2020

VA purchased two types of eye tracking equipment for veterans.

- VA reported it purchased prosthetics that enable veterans with speech impairment or loss to communicate using a computer or tablet device. VA reported purchasing 42 devices for $416,284.
- VA reported it purchased a prosthetic device that enabled one veteran to use eye gaze to manipulate a laptop computer for $18,860.

Regulation of Nonfederal Entities' Use of FRT in Fiscal Year 2020

None reported.

Planned Use of FRT Systems through Fiscal Year 2023

VA reported it plans to use two FRT systems through fiscal year 2023, for physical security and domestic law enforcement purposes.

Department of Veterans Affairs (VA) Plans for Facial Recognition Technology (FRT) Systems through Fiscal Year 2023		
FRT system	**Description of system use**	**Planned new use status**
New federal FRT systems		
VA Police Service Chicago, IL Motorola Avigilon	The FRT system can sort through video quickly locating a specific person, such as a missing patient or to input a photo from a police bulletin (e.g., 'Be On the Lookout'), to alert officers when that person enters the VA medical center campus.	VA reported that they plan to obligate funding for a project design to install Motorola Avigilon by fiscal year 2022.
VA Police Service West Palm Beach, FL Veritone ai Ware	The web-based FRT can detect and follow selected, moving objects, such as a previously disruptive individual or to track missing patients, and will alert officers when on the VA medical center campus.	VA purchased FRT software for 12 cameras in 2019, and is working with the contractor to ensure the system is operational by the middle of fiscal year 2022.

Source: GAO analysis of survey results. | GAO-21-526

Components with No FRT-Related Activities in Fiscal Year 2020 or Planned Uses through Fiscal Year 2023

According to VA, the following components reported no FRT-related activities in fiscal year 2020 or planned uses through fiscal year 2023 listed above:

- Veterans Benefits Administration
- National Cemetery Administration

GSA

The General Services Administration (GSA) reported facial recognition technology (FRT) activities in fiscal year 2020. Specifically, GSA reported owning one FRT system. GSA reported it does not plan to use other FRT systems through fiscal year 2023.

Use of FRT Systems in Fiscal Year 2020

GSA reported it owned one FRT system in fiscal year 2020, for digital access purposes.

General Services Administration (GSA) Facial Recognition Technology (FRT) Systems Used in Fiscal Year 2020		
FRT system	**Description of system use**	**Other GSA and federal users**
GSA-owned federal FRT system		
Technology Transformation Service login.gov	Login.gov conducted a pilot with Department of the Treasury employees to test FRT services that compared two photos to verify the identity of an individual accessing the website or application. Login.gov takes a picture of the individual and their photo identification to determine a match.	• Department of the Treasury

Source: GAO analysis of survey results. | GAO-21-526

FRT System Obligations in Fiscal Year 2020

GSA reported it obligated over $90,000 for login.gov in fiscal year 2020.

Research & Development in Fiscal Year 2020

None reported.

Transactions with Nonfederal Entities in Fiscal Year 2020

None reported.

Regulation of Nonfederal Entities' Use of FRT in Fiscal Year 2020

None reported.

Planned Use of FRT Systems through Fiscal Year 2023

None reported.

Components with No FRT-Related Activities in Fiscal Year 2020 or Planned Uses through Fiscal Year 2023

According to GSA, the following components reported no FRT-related activities in fiscal year 2020 or planned uses through fiscal year 2023 listed above:

- National Services (Public Building Service)
- Regional Offices
- Staff Offices
- Independent Offices (Office of Inspector General)

The National Aeronautics and Space Administration (NASA) reported facial recognition technology (FRT) activities in fiscal year 2020. Specifically, NASA reported testing one FRT system and conducting FRT-related research and development (R&D). NASA reported it does not plan to use other FRT systems through fiscal year 2023.

Use of FRT Systems in Fiscal Year 2020

NASA reported it owned one FRT prototype system in fiscal year 2020. Specifically, NASA tested this system for issuing badges.

National Aeronautics and Space Administration (NASA) Facial Recognition Technology (FRT) Systems Used in Fiscal Year 2020

FRT system	Description of system use	Other NASA and federal users
NASA-owned FRT system		
Johnson Space Center Forgotten badge prototype	The forgotten badge prototype confirms an employee's identity by comparing a current camera image of the employee with a photo on file. NASA reported that it would not continue work on the prototype due to cost	• None

Source: GAO analysis of survey results. | GAO-21-526

FRT System Obligations in Fiscal Year 2020

NASA reported it did not obligate funds for its forgotten badge prototype in fiscal year 2020, because funding for the prototype was obligated in fiscal year 2019.

Research & Development in Fiscal Year 2020

NASA reported using FRT as a tool to conduct human factors research. Specifically, the Langley Research Center conducted a series of controlled research experiments from 2013 to 2020, to understand the cognitive states of aircraft and space flight crew. These experiments included the use of eye tracking and facial muscle tracking devices to understand cognitive states (e.g., surprised, focused) and other human factors during simulations. NASA reported it obligated $60,000 for the R&D experiments in fiscal year 2020.

Transactions with Nonfederal Entities in Fiscal Year 2020

None reported.

Regulation of Nonfederal Entities' Use of FRT in Fiscal Year 2020

None reported.

Planned Use of FRT Systems through Fiscal Year 2023

None reported.

Components with No FRT-Related Activities in Fiscal Year 2020 or Planned Uses through Fiscal Year 2023

According to NASA, the following components reported no FRT-related activities in fiscal year 2020 or planned uses through fiscal year 2023 listed above:

- Administrator Staff Offices
- Mission Directorates
- Mission Support Directorate
- Office of the Administrator
- Office of the Inspector General

The National Science Foundation (NSF) reported facial recognition technology (FRT) activities in fiscal year 2020. Specifically, NSF reported it supported FRT-related research and development (R&D) in fiscal year 2020. NSF also reported it does not plan to use other FRT systems through fiscal year 2023.

Use of FRT Systems in Fiscal Year 2020

None reported.

FRT System Obligations in Fiscal Year 2020

None reported.

Research & Development in Fiscal Year 2020

NSF reported it supported FRT-related R&D conducted by external organizations in fiscal year 2020. Specifically, NSF's Directorate for Computer and Information Science and Engineering awarded three grants to universities and others to conduct research on FRT and related areas. For example:

- The University of Chicago conducted research using machine learning for facial recognition.
- The Columbus State University conducted research on facial privacy.
- The University of Nebraska conducted research using eye tracking to optimize programmer productivity.

NSF reported obligating $1,696,146 for these grants in fiscal year 2020, but this amount includes areas of research other than FRT.

Transactions with Nonfederal Entities in Fiscal Year 2020

None reported.

Regulation of Nonfederal Entities' Use of FRT in Fiscal Year 2020

None reported.

Planned Use of FRT Systems through Fiscal Year 2023

None reported.

Components with No FRT-Related Activities in Fiscal Year 2020 or Planned Uses through Fiscal Year 2023

According to NSF, the following components reported no FRT-related activities in fiscal year 2020 or planned uses through fiscal year 2023 listed above:

- Chief Information Officer
- Directorate for Biological Sciences
- Directorate for Education and Human Resources
- Directorate for Engineering

- Directorate for Geosciences
- Directorate for Mathematical and Physical Sciences
- Directorate for Social, Behavioral, and Economic Sciences
- National Science Board (Office of Inspector General)
- Office of Budget, Finance, and Award Management
- Office of the Director
- Office of Diversity and Inclusion
- Office of the General Counsel
- Office of Information Resource Management
- Office of Integrative Activities
- Office of International Science and Engineering
- Office of Legislative and Public Affairs

The Social Security Administration (SSA) reported facial recognition technology (FRT) activities in fiscal year 2020. Specifically, SSA reported accessing one FRT system in fiscal year 2020. SSA reported it does not plan to use other FRT systems through fiscal year 2023.

Use of FRT Systems in Fiscal Year 2020

SSA reported it accessed one FRT system in fiscal year 2020, for digital access purposes.

Social Security Administration (SSA) Facial Recognition Technology (FRT) Systems Used in Fiscal Year 2020

FRT system	Description of system use	Other SSA and federal users
Accessed commercial FRT system		**SSA users**
Acuant FaceID	This pilot of Acuant FaceID was conducted using agency employees to verify the identity of employees within the test group for access to SSA's public online services by remotely confirming that the facial image on an identity document (e.g., state-issued IDs) matches the facial image of the applicant.	• Office of Digital Transformation

Source: GAO analysis of survey results. | GAO-21-526

FRT System Obligations in Fiscal Year 2020

SSA reported that it could not provide the specific amount that was obligated for FRT, because this FRT was included as part of a larger task order. Instead, SSA reported that it obligated $809,179 as part of a larger contract to support digital identity enhancements, including the pilot.

Research & Development in Fiscal Year 2020

None reported.

Transactions with Nonfederal Entities in Fiscal Year 2020

None reported.

Regulation of Nonfederal Entities' Use of FRT in Fiscal Year 2020

None reported.

Planned Use of FRT Systems through Fiscal Year 2023

None reported.

Components with No FRT-Related Activities in Fiscal Year 2020 or Planned Uses through Fiscal Year 2023

According to SSA, the following components reported no FRT-related activities in fiscal year 2020 or planned uses through fiscal year 2023 listed above:

- Office of Analytics, Review, and Oversight
- Office of Budget, Finance, and Management
- Office of the Chief Actuary
- Office of the Commissioner
- Office of Communications
- Office of General Counsel
- Office of the Inspector General
- Office of Hearings Operations
- Office of Human Resources
- Office of Legislation and Congressional Affairs
- Office of Operations
- Office of Retirement and Disability Policy

Appendix III : Comments from the U.S. Agency for International Development

USAID
FROM THE AMERICAN PEOPLE

June 30, 2021

Candice N. Wright
Acting Director
Science, Technology Assessment, and Analytics

and

Gretta L. Goodwin
Director
Homeland Security and Justice

U.S. Government Accountability Office
441 G Street, N.W.
Washington, D.C. 20226

Re: GAO draft report entitled "Facial Recognition Technology: Current and Planned Uses by Federal Agencies". GAO-21-526SU (Engagement Code 104231)

Dear Ms. Wright and Ms. Goodwin:

I am pleased to provide the formal response of the U.S. Agency for International Development (USAID) to the draft report produced by the U.S. Government Accountability Office (GAO) titled, "Facial Recognition Technology: Current and Planned Uses by Federal Agencies", GAO-21-526SU (Engagement Code 104231)

The report has no recommendations for USAID and we do not offer specific comments with regards to content of the report. The Agency continues to apply fiscal responsibility with regard to information technology investments, including emerging technologies. This is evidenced in our four overall "A" ratings on Federal Information Technology Acquisition Reform Act (FITARA) Scorecards, including for incremental development, transparency and risk management, the consolidation and optimization of our data centers, and software licensing.

The Agency is continuously committed to identifying and delivering efficient, flexible information technology solutions. This is evident in our current use of facial recognition technology (FRT) for digital access on our mobile devices, which provides high value to our workforce, without compromising security and privacy of our data. USAID will surely monitor the advancement of FRT in the coming years and assess how and when further adoption is appropriate or applicable to meet the Agency's needs and foster partnership.

Thank you for the opportunity to respond to the draft report, and for the courtesies extended by your staff while conducting this engagement. We appreciate the opportunity to participate in the complete and thorough evaluation of our Facial Recognition initiative.

Sincerely,

Colleen R. Allen

Colleen R. Allen
Acting Assistant Administrator
Bureau for Management

Appendix IV : Comments from the Social Security Administration

SOCIAL SECURITY
Office of the Commissioner

July 6, 2021

Candice N. Wright
Director, Science, Technology Assessment, and Analytics
U.S. Government Accountability Office
441 G Street, NW
Washington, DC 20548

Dear Director Wright:

Thank you for the opportunity to review the draft report "FACIAL RECOGNITION TECHNOLOGY: Current and Planned Uses by Federal Agencies" (GAO-21-526SU). The report is an accurate assessment of our experience related to facial recognition technology.

If you have any questions, please contact Trae Sommer, Director of the Audit Liaison Staff, at (410) 965-9102.

Sincerely,

Scott Frey
Chief of Staff

SOCIAL SECURITY ADMINISTRATION BALTIMORE, MD 21235-0001

Chapter 3

Facial Recognition Technology: Federal Law Enforcement Agencies Should Have Better Awareness of Systems Used by Employees*

Gretta L. Goodwin

Why GAO Did This Study

Federal agencies that employ law enforcement officers use facial recognition technology to assist criminal investigations, among other activities. For example, the technology can help identify an unknown individual in a photo or video surveillance.

This statement describes (1) the ownership and use of facial recognition technology by federal agencies that employ law enforcement officers, (2) the types of activities these agencies use the technology to support, and (3) the extent that these agencies track employee use of facial recognition technology owned by non-federal entities, including the potential privacy and accuracy implications.

This statement is based on GAO's June 2021 report on federal law enforcement's use of facial recognition technology (GAO-21-518). To conduct that prior work, GAO administered a survey questionnaire to 42

* This is an edited, reformatted and augmented version of the Governmental Accountability Office Testimony before the Subcommittee on Crime, Terrorism, and Homeland Security, Committee on the Judiciary, House of Representatives, Publication No. GAO-21-105309, dated July 13, 2021.

In: Facial Recognition Technology
Editor: Mari F. Burke
ISBN: 979-8-88697-124-8

federal agencies that employ law enforcement officers regarding their use of the technology. GAO also reviewed relevant documents and interviewed agency officials. The June 2021 report was a public version of a sensitive report that GAO issued in April 2021. Information that agencies deemed sensitive was omitted from the June 2021 report and this statement.

What GAO Recommends

In June 2021, GAO made two recommendations to each of 13 federal agencies to implement a mechanism to track what non-federal systems are used by employees, and assess the risks of using these systems. Agencies generally concurred with the recommendations.

What GAO Found

In June 2021, GAO reported the results of its survey of 42 federal agencies that employ law enforcement officers about their use of facial recognition technology. Twenty reported owning systems with the technology or using systems owned by other entities, such as state, local, and non-government entities (see figure).

Agencies reported using the technology to support several activities (e.g., criminal investigations) and in response to COVID-19 (e.g., verify an individual's identity remotely). Six agencies reported using the technology on images of the unrest, riots, or protests following the killing of Mr. George Floyd in May 2020. Three agencies reported using it on images of the U.S. Capitol attack on January 6, 2021. Agencies said the searches used images of suspected criminal activity.

Fourteen of the 42 agencies reported using the technology to support criminal investigations. However, only one had a mechanism to track what non-federal systems were used by employees. By having a mechanism to track use of these systems and assessing the related risks (e.g., privacy and accuracy-related risks), agencies can better mitigate risks to themselves and the public.

Chair Jackson Lee, Ranking Member Biggs, and Members of the Subcommittee:

I am pleased to be here to discuss federal law enforcement's use of facial recognition technology. Use of this technology has expanded in recent years, raising concerns about privacy and the accuracy of facial recognition systems. Members of Congress and academics have highlighted the importance of understanding what technologies are owned and how they are used by federal law enforcement.

My statement today will focus on (1) the ownership and use of facial recognition technology by federal agencies that employ law enforcement officers, (2) the types of activities these agencies use the technology to support, and (3) the extent that these agencies track employee use of facial recognition technology owned by non-federal entities, including the potential privacy and accuracy implications. This statement is based on findings from our June 2021 report, which included the results of our survey questionnaire to 42 federal agencies that employ law enforcement officers about their use of the technology.[1] The report provides a detailed description of our scope and methodology.

The work upon which this statement is ba"ed was cond"cted in accordance with generally accepted government auditing standards. Those standards require that we plan and perform the audit to obtain sufficient, appropriate evidence to provide a reasonable basis for our findings and conclusions based on our audit objectives. We believe that the evidence obtained provides a reasonable basis for our findings and conclusions based on our audit objectives.

Background

Federal and Non-Federal Systems with Facial Recognition Technology

Federal law enforcement may use systems with facial recognition technology owned by their respective agencies. They may also use systems owned by other government entities, including federal, state, local, tribal, and territorial government entities. Moreover, federal law enforcement may use non-

[1] 1GAO, Facial Recognition Technology: Federal Law Enforcement Agencies Should Better Assess Privacy and Other Risks, GAO-21-518 (Washington D.C.: June 3, 2021). Our June 2021 report was a public version of a sensitive report that we issued in April 2021 (GAO-21-243SU). Some federal agencies deemed information in our April 2021 report to be sensitive, which must be protected from public disclosure. Therefore, the June 2021 report and this statement omit sensitive information about federal agency ownership and use of facial recognition technology.

government facial recognition service providers, such as Vigilant Solutions and Clearview AI. For example, law enforcement officers with a Clearview AI account may use a computer or smartphone to upload a photo of an unknown individual to Clearview AI's facial recognition system. The system can return search results that show potential photos of the unknown individual, as well as links to the site where the photos were obtained (e.g., Facebook).

In some cases, law enforcement officers can access another entity's system and conduct a facial recognition search. Alternatively, law enforcement officers can request that another entity use its system to conduct facial recognition searches on their behalf. For example, federal law enforcement officers may ask a state police department to conduct facial recognition searches on their behalf.

Privacy Laws and Rules

Several statutory requirements govern the protection of personal information by federal agencies, including federal law enforcement's use of facial images. The Privacy Act of 1974 places limitations on agencies' collection, disclosure, and use of personal information maintained in systems of records (e.g., photos). According to Office of Management and Budget (OMB) officials, the Privacy Act and OMB Circular A-130 generally provide that agencies must ensure that privacy requirements apply to systems operated by contractors or other entities on behalf of the Federal Government, which could include facial recognition service providers.

Accuracy of Facial Recognition Technology

The accuracy of facial recognition technology can be characterized in a number of ways. For example, a false positive rate is how often the technology incorrectly declares two images to be a match when they are actually from two different people. In addition, a false negative rate is how often the technology fails to declare two images "to be a match " when they are actually from the same person. Matching errors can be caused not only by the quality of the facial recognition technology, but also by the quality of the photos used in the matching process and other factors. The National Institute of Standards and Technology has conducted research into the accuracy of facial recognition algorithms. It has evaluated hundreds of commercial facial matching algorithms for accuracy and speed since 2000.

Twenty Federal Agencies Reported Owning or Using Systems with Facial Recognition Technology

We surveyed 42 federal agencies that employ law enforcement officers, 20 of which reported that they owned a system with facial recognition technology or used another entity's system.[2] As shown in figure 1, three agencies only owned a system, 12 agencies only used another entity's system, and five agencies both owned a system and used another entity's system. According to these agencies, some systems can include hundreds of millions or billions of photos.

Eight Agencies Reported Owning Systems with Facial Recognition Technology

Eight of the 42 federal agencies reported owning 17 systems with facial recognition technology, from January 2015 through March 2020.[3] Of these systems, four were in operation as of March 31, 2020, and were owned by three agencies: the Federal Bureau of Investigation, Federal Bureau of Prisons, and U.S. Customs and Border Protection. In addition, one of the eight agencies reported that it was in the process of procuring two systems during this time

[2] Specifically, we asked agencies whether at any point from January 2015 through March 2020, they owned a system with facial recognition technology, including systems in the process of being developed. In addition, we asked agencies whether at any point from April 2018 through March 2020, they used facial recognition technology—that is, their offices, employees, or contractors (1) accessed a system owned/operated by another entity, or (2) requested that another entity use its system to conduct a facial recognition search on their behalf. See the complete list of 42 federal agencies that received our questionnaire in appendix I.

We defined facial recognition technology as a type of automated or semi-automated biometric technology that uses images for verification, identification, and/or investigative purposes. In addition, we stated that a system with facial recognition technology may include a facial recognition algorithm, hardware, software, and a photo database. We asked agencies to include all uses of facial recognition technology in their response except for facial recognition technology that was solely used to authenticate the identity of the agency's employees and contractors to log into computers and phones.

[3] This statement omits some information about systems owned by agencies we surveyed, as the relevant agencies deemed the information sensitive.

period, but had not finalized the purchase as of March 2020. Detailed descriptions of the 19 systems (17 owned and two in procurement) and their status as of March 31, 2020 can be found in our June 2021 report.

Source: GAO analysis of survey data. | GAO-21-15309.

Note: We sent a survey questionnaire to 42 federal agencies that employ law enforcement officers. We asked agencies whether at any point during January 2015 through March 2020, they owned a system with facial recognition technology, including systems in the process of being developed. In addition, we asked agencies whether at any point from April 2018 through March 2020, they used facial recognition technology—that is, their offices, employees, or contractors (1) accessed a system owned or operated by another entity, or (2) requested that another entity use its system to conduct a facial recognition search on their behalf.

The owned system columns include systems in the process of being developed. The National Aeronautics and Space Administration's Office of Protective Services reported that it did not purchase facial recognition technology. However, we included the agency in the owned column because it used a commercial-off-the-shelf product with facial recognition technology to conduct a proof of concept test to determine whether the technology was suitable for its purposes.

Figure 1. Ownership and Use of Facial Recognition Technology Reported by Federal Agencies that Employ Law Enforcement Officers.

Seventeen Federal Agencies Reported Using Systems Owned by Other Entities

Seventeen of the 42 federal agencies reported using another entity's system with facial recognition technology from April 2018 through March 2020. Of the 17 agencies, 15 reported using systems owned by another federal entity; 14 reported using systems owned by state, local, tribal, or territorial entities; and 11 reported using systems owned by non- government entities. Furthermore, nine of the 17 agencies reported using systems owned by all three types of entities. See table 1 for additional information.

Table 1. Reported use of other entities' facial recognition technology by federal agencies that employ law enforcement officers

Type of Entity That Owned System					
			Non-Government[a]		
Federal Agency That Used System	Other Federal	State, Local, Tribal, Territorial	Clear-view AI	Vigilant Solutions	Other Non-Govern-ment[b]
Bureau of Diplomatic Security	✓	✓	✓	✓	✓
U.S. Customs and Border Protection	✓	✓	✓	✓	✓
U.S. Marshals Service	✓	✓	✓	✓	✓
Bureau of Alcohol, Tobacco, Firearms and Explosives	✓	✓	✓	✓	—
U.S. Immigration and Customs Enforcement	✓	✓	✓	—	✓
U.S. Postal Inspection Service	✓	✓	✓	✓	—
Drug Enforcement Administration	✓	✓	✓	—	—
Federal Bureau of Investigation	✓	✓	✓	—	—
U.S. Secret Service	✓	✓	✓	—	—
U.S. Capitol Police	✓	✓	—	—	—
U.S. Fish and Wildlife Service	✓	✓	—	—	—
Food and Drug Administration, Office of Criminal Investigations	✓	✓	—	—	—

Table 1. Continued

Type of Entity That Owned System					
			Non-Government[a]		
Federal Agency That Used System	Other Federal	State, Local, Tribal, Territorial	Clear-view AI	Vigilant Solutions	Other Non-Govern-ment[b]
Internal Revenue Service, Criminal Investigation Division	✓	✓	—	—	—
U.S. Park Police	—	✓	✓	—	—
Administrative Office of the U.S. Courts, U.S. Probation and Pretrial Services	—	—	—	—	✓
Pentagon Force Protection Agency	✓	—	—	—	—
Transportation Security Administration	✓	—	—	—	—
Total	15	14	10	5	5

Source: GAO analysis of survey data. I GAO-21-105309.

Legend:

✓ Agency used a system owned by the respective entity (or entity type) at any point from April 2018 through March 2020. For federal, state, local, tribal, and territorial entities, the term "used" includes an agency's offices, employees, or contractors (1) accessing a system owned/operated by the respective entity type, or (2) requesting that the respective entity type use its system to conduct a facial recognition search on the agency's behalf. For non- government entities, the term "used" means the agency's offices, employees, or contractors submitted photos to the respective non-government service provider for the purpose of conducting a facial recognition search.

– Agency did not use a system owned by the respective entity (or entity type) at any point from April 2018 through March 2020.

[a]Some agencies reported that they only used Clearview AI or Vigilant Solutions on a free trial basis, and thus, did not enter into a formal contract with the service provider.

[b]Other non-government entities that agencies reported using included Amazon Rekognition, BI SmartLink, and Giant Oak Social Technology, among others.

Agencies Reported Using Systems that Varied in the Number and Type of Photos

Federal agencies reported using numerous systems with facial recognition technology, and sometimes these systems included stored photos. The number and types of photos within these systems can vary, based on information reported by agencies and system owners. Reported photo types included: mug shot photos, driver's license photos, passport photos, publicly available photos on the internet, and images from video/Closed Circuit Television. Figure 2 below shows examples of federal, state, and non-government systems with facial recognition technology that federal agencies reported using, and the number of photos in them.

Federal Agencies Reported Using Systems with Facial Recognition Technology to Support Various Activities

Federal agencies reported using facial recognition technology to support various activities, such as criminal investigations and surveillance, and also in response to the Coronavirus Disease 2019 (e.g., verify an individual's identity remotely). Of the 20 agencies that owned or used facial recognition technology, 14 reported using the technology to support criminal investigations. For example, the FBI's Next Generation Identification Interstate Photo System allows users to search a database of over 40 million photos. The system returns a list of potential candidates that law enforcement can use to generate investigative leads. According to the FBI, the system has been used for investigations of violent crimes, credit card and identity fraud, missing persons, and bank robberies, among others.

Six agencies reported using facial recognition technology during May through August 2020 to support criminal investigations related to civil unrest, riots, or protests.[4] Following the killing of Mr. George Floyd while in the custody of the Minneapolis, Minnesota police department on May 25, 2020, nationwide civil unrest, riots, and protests occurred. Six agencies told us that they used images from these events to conduct facial recognition searches during May through August 2020 in order to assist with criminal investigations (see table 2). All six agencies reported that these searches were on images of individuals suspected of violating the law.

---

[4] We requested this information from 17 agencies that indicated in their questionnaire response as (1) having a system with facial recognition technology that was in operation, or (2) using another entity's system.

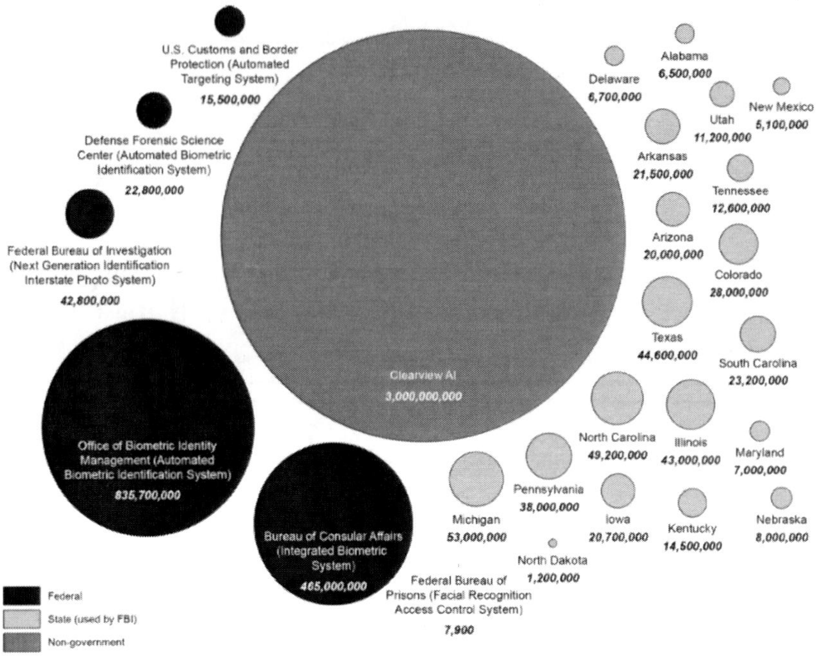

Source: GAO analysis of information provided by system users or owners. | GAO-21-105309.

Note: We sent a survey questionnaire to 42 federal agencies that employ law enforcement officers. This figure includes examples of systems used by one or more federal agencies we surveyed. It does not include all systems used by these agencies. The figure includes the number of photos stored in the respective entity's system with facial recognition technology, as of March 31, 2020.

The same individual may be included in multiple photos within one photo database or across multiple databases, and the same photo can exist within multiple databases. Some entities providing these numbers indicated they were estimates. The number of photos for federal and non-government entities were reported by the respective system owner. The number of photos for state entities were all reported by the Federal Bureau of Investigation (FBI). Specifically, the FBI's Facial Analysis, Comparison, and Evaluation Services has memorandums of understanding with several state agencies, allowing it to leverage the state-owned systems for facial recognition searches. The FBI provided the number of photos they can access via these memorandums of understanding.

Figure 2. Selected Federal, State, and Non-government Systems with Facial Recognition Technology Used by Federal Agencies that Employ Law Enforcement Officers, and the Number of Photos in Them.

Three agencies reported using facial recognition technology on images from the U.S. Capitol attack on January 6, 2021. The three agencies reported using the technology to generate investigative leads for criminal investigations as follows:

- U.S. Capitol Police used Clearview AI to help generate investigative leads. The agency also requested that another federal agency use its system to conduct facial recognition searches on behalf of the U.S. Capitol Police.
- U.S. Customs and Border Protection used its Automated Targeting System to conduct searches at the request of another federal agency.
- Bureau of Diplomatic Security used the Department of State's Integrated Biometric System to conduct searches at the request of another federal agency.

Agencies also reported using facial recognition technology to support other activities, such as surveillance, traveler verification, area access, and research and education. For example, the U.S. Secret Service piloted a system with facial recognition technology to determine whether it could be incorporated into the agency's White House Complex security operations. Specifically, the Secret Service stored photos of 23 volunteer employees within the system. As volunteers moved throughout the White House Complex, their images were captured by closed-circuit television cameras. In real time, the system compared the stored photos to images from the video footage to determine whether they represented the same individual. Secret Service told us it did not plan to implement the system based on the results of the pilot.

Table 2. Federal agency reported use of facial recognition technology on images of individuals suspected of violating the law during civil unrest, riots, or protests, May through August 2020

Federal Agency	How Agency Reported Using Facial Recognition Technology
Bureau of Alcohol, Tobacco, Firearms and Explosives	In a single instance, used facial recognition technology owned by another law enforcement entity. The search was conducted to help identify an individual suspected of violating the law during the period of civil unrest, riots, or protests.

Federal Agency	How Agency Reported Using Facial Recognition Technology
U.S. Capitol Police	Requested that the Montgomery County Department of Police (Montgomery County, Maryland) conduct facial recognition searches to assist with a criminal investigation. The purpose of the searches was to help identify individuals that confronted and made threats to a member of Congress and the member's spouse outside the White House during the period of civil unrest, riots, or protests.
Federal Bureau of Investigation	Created a digital media tip line and solicited images of people involved in criminal activity during the period of civil unrest, riots, or protests. The agency sought to identify or locate criminal suspects seen in images and video depicting criminal behavior by conducting facial recognition searches using its Next Generation Identification Interstate Photo System.
U.S. Marshals Service	Used a non-government facial recognition service provider to conduct facial recognition searches related to criminal investigations on images from the period of civil unrest, riots, or protests.
U.S. Park Police	Requested that the Maryland National Capital Park Police conduct a facial recognition search using an image from Twitter to identify an individual who allegedly assaulted an officer during the period of civil unrest, riots, or protests. The search was conducted on the National Capital Region Facial Recognition Investigative Leads System. The subject was ultimately charged with Felony Civil Disorder and two counts of Assault on a Police Officer.
Federal Agency	How Agency Reported Using Facial Recognition Technology
U.S. Postal Inspection Service	Used Clearview AI to help identify individuals suspected of criminal activity that took place in conjunction with the period of civil unrest, riots, or protests. This criminal activity included damaging U.S. Postal Service property, stealing mail, opening mail, burglarizing U.S. Postal Service buildings, and committing arson.

Source: GAO analysis of survey data I GAO-21-105309.

Most Agencies Do Not Track Non- Federal Systems in Use or Related Risks

Thirteen agencies do not have complete, up-to-date information on what non-federal systems are used by employees.[5] These 13 agencies have therefore not fully assessed the potential risks of using these systems, such as risks related to privacy and accuracy. Most federal agencies that reported using non-federal systems did not own systems. Thus, employees were relying on systems owned by other entities, including non-federal entities, to support their operations.

Specifically, we found that 13 of 14 agencies that reported using non-federal systems do not have a mechanism to track what non-federal systems are used by employees (see table 3).[6] For example, when we requested information from one of the agencies about its use of non- federal systems, agency officials told us they had to poll field division personnel because the information was not maintained by the agency. These agency officials also told us that the field division personnel had to work from their memory about their past use of non-federal systems, and that they could not ensure we were provided comprehensive information about the agency's use of non-federal systems. Officials from another agency initially told us that its employees did not use non-federal systems; however, after conducting a poll, the agency learned that its employees had used a non-federal system to conduct more than 1,000 facial recognition searches.

Numerous risks to federal agencies and the public can accompany the use of facial recognition technology. In particular, these risks can relate to privacy and the accuracy of a system. For example, when agencies use facial recognition technology without first assessing the privacy implications, there is a risk that the agencies will not adhere to privacy- related laws, regulations, and policies. There is also a risk that non- federal system owners will share sensitive information (e.g., photo of a suspect) about an ongoing investigation with the public or others.

---

[5] By complete, up-to-date information, we mean that an agency has ongoing knowledge of what non-federal systems with facial recognition technology are used by employees. By non-federal systems, we are referring to systems owned by state, local, tribal, territorial, and non-government entities.

[6] Fifteen agencies reported using non-federal systems; however, we excluded U.S. Probation and Pretrial Services because it does not use facial recognition technology to support criminal investigations. All 14 agencies discussed in this section reported using the technology to support criminal investigations.

Table 3. Tracking of employee use of non-federal systems with facial recognition technology among selected federal agencies

Federal Agency	Have Mechanism to Track What Non-Federal Systems Are used by Employees
U.S. Immigration and Customs Enforcement	Yes
Bureau of Alcohol, Tobacco, Firearms and Explosives	No
Bureau of Diplomatic Security	No
U.S. Capitol Police	No
U.S. Customs and Border Protection	No
Drug Enforcement Administration	No
Federal Bureau of Investigation	No
U.S. Fish and Wildlife Service	No
Food and Drug Administration, Office of Criminal Investigations	No
Internal Revenue Service, Criminal Investigation Division	No
U.S. Marshals Service	No
U.S. Park Police	No
U.S. Postal Inspection Service	No
U.S. Secret Service	No

Source: GAO analysis of agency information. I GAO-21-105309.

Note: Federal agencies marked "No" may have known that employees used certain systems, but they do not have a mechanism to provide complete, up-to-date information of what systems are used by employees.

Although the accuracy of facial recognition technology has increased dramatically in recent years, risks still exist that searches will provide inaccurate results. For example, if a system is not sufficiently accurate, it could unnecessarily identify innocent people as investigative leads. The system could also miss investigative leads that could otherwise have been revealed. In December 2019, the National Institute of Standards and Technology reported that facial recognition algorithms it tested differed in accuracy widely by race, ethnicity, or country of origin, as well as by gender and age.[7] Some members of Congress, privacy groups, and others have expressed concerns

[7] National Institute of Standards and Technology, Face Recognition Vendor Test (FRVT) Part 3: Demographic Effects, NIST Interagency or Internal Report 8280 (Dec. 19, 2019). The National Institute of Standards and Technology reported that it tested 189 mostly commercial algorithms from 99 developers and that performance differences varied by the algorithms tested, with some performing better than others. For a small number of the one-to-many algorithms, differences in false positives across demographic groups were undetectable. The extent of performance differences varied by the developer, type of error, and quality of the facial images.

that facial recognition technology's higher error rates for certain demographics could result in disparate treatment, profiling, or other adverse consequences for members of these populations.

One agency—the U.S. Immigration and Customs Enforcement—reported that it was in the process of implementing a mechanism to track what non-federal systems are used by employees. According to U.S. Immigration and Customs Enforcement officials, in November 2020 they were in the process of developing a list of approved facial recognition technologies that employees can use. In addition, log-in sheets will be made available to employees, allowing supervisors to monitor employee use of the technologies. The agency also assessed privacy and accuracy risks associated with its use of facial recognition technology, including non-federal systems.[8]

However, 13 federal agencies cannot fully assess the risks of using non-federal systems because they do not have complete, up-to-date information on what systems are actually used by employees. Therefore, in June 2021 we recommended that these 13 agencies: (1) implement a mechanism to track what non-federal systems with facial recognition technology are used by employees to support investigative activities; and (2) after implementing a mechanism to track non-federal systems, assess the risks of using such systems, including privacy and accuracy-related risks.[9]

These agencies generally concurred with our recommendations.[10] We believe that expeditious implementation of our recommendations is essential. By implementing a mechanism to track what non-federal systems are used by employees, agencies will have better visibility into the technologies they rely upon to conduct criminal investigations. In addition, by assessing the risks of using these systems, including privacy and accuracy-related risks, agencies will be better positioned to mitigate any risks to themselves and the public.

Chair Jackson Lee, Ranking Member Biggs, and Members of the Subcommittee, this concludes my prepared statement. I would be pleased to respond to any questions you may have at this time.

---

[8] Department of Homeland Security, Privacy Impact Assessment for the ICE Use of Facial Recognition Services, DHS/ICE/PIA-054 (May 13, 2020).

[9] We made these recommendations to all 13 agencies that reported "No" in table 3 above.

[10] As of June 2021, the 13 federal agencies have not implemented these recommendations.

Appendix I: Federal Agencies That Participated in GAO's Survey Questionnaire on Facial Recognition Technology

Our June 2021 report included the results of a survey questionnaire we sent to 42 federal agencies that employed law enforcement officers.[11] The survey focused on agency ownership and use of facial recognition technology. See the list of 42 federal agencies in table 4 below.

Table 4. Federal agencies that participated in GAO's Survey Questionnaire on facial recognition technology

Federal Agency	Department
Forest Service	Agriculture
Bureau of Industry and Security	Commerce
National Oceanic and Atmospheric Administration, Office of Law Enforcement	Commerce
Office of Security	Commerce
Secretary's Protective Detail	Commerce
Pentagon Force Protection Agency	Defense
National Nuclear Security Administration	Energy
Food and Drug Administration, Office of Criminal Investigations	Health and Human Services
National Institutes of Health, Division of Police	Health and Human Services
U.S. Customs and Border Protection	Homeland Security
Federal Emergency Management Agency, Mount Weather Police	Homeland Security
Federal Protective Service	Homeland Security
U.S. Immigration and Customs Enforcement	Homeland Security
Office of the Chief Security Officer	Homeland Security
U.S. Secret Service	Homeland Security
Transportation Security Administration	Homeland Security
Bureau of Indian Affairs, Office of Justice Services	Interior
Bureau of Land Management	Interior
Bureau of Reclamation	Interior
U.S. Fish and Wildlife Service	Interior
U.S. Park Police	Interior
National Park Service Rangers	Interior
Bureau of Alcohol, Tobacco, Firearms, and Explosives	Justice
Drug Enforcement Administration	Justice
Federal Bureau of Investigation	Justice

[11] *GAO, Facial Recognition Technology: Federal Law Enforcement Agencies Should Better Assess Privacy and Other Risks,* GAO-21-518 (Washington D.C.: June 3, 2021). In our June 2021 report, we defined federal law enforcement officers as full-time employees with federal arrest authority and who are authorized to carry firearms while on duty. See more information on our scope and methodology in our June 2021 report.

Federal Bureau of Prisons	Justice
U.S. Marshals Service	Justice
Division of Protective Operations	Labor
Bureau of Diplomatic Security	State
Federal Agency	Department
Bureau of Engraving and Printing Police	Treasury
Internal Revenue Service, Criminal Investigation Division	Treasury
U.S. Mint Police	Treasury
Police Service	Veterans Affairs
Amtrak Police Department	–
Environmental Protection Agency, Criminal Investigation Division	–
National Aeronautics and Space Administration, Office of Protective Services	–
U.S. Postal Inspection Service	–
Smithsonian Institution, Office of Protection Services	–
Tennessee Valley Authority Police	–
Administrative Office of the U.S. Courts, U.S. Probation and Pretrial Services	–
U.S. Capitol Police	–
Government Publishing Office, Uniform Police Branch	–

Source: GAO information. I GAO-21-105309.

Legend: – Not Applicable.

Chapter 4

Facial Recognition Technology: Federal Law Enforcement Agencies Should Better Assess Privacy and Other Risks*

United States Government Accountability Office

Why GAO Did This Study

Federal agencies that employ law enforcement officers can use facial recognition technology to assist criminal investigations, among other activities. For example, the technology can help identify an unknown individual in a photo or video surveillance.

GAO was asked to review federal law enforcement use of facial recognition technology. This chapter examines the 1) ownership and use of facial recognition technology by federal agencies that employ law enforcement officers, 2) types of activities these agencies use the technology to support, and 3) the extent that these agencies track employee use of facial recognition technology owned by non-federal entities.

GAO administered a survey questionnaire to 42 federal agencies that employ law enforcement officers regarding their use of the technology. GAO also reviewed documents (e.g., system descriptions) and interviewed officials from selected agencies (e.g., agencies that owned facial recognition

* This is an edited, reformatted and augmented version of United States Government Accountability Office, Report to Congressional Requesters, Publication No. GAO-21-518, dated June 2021.

In: Facial Recognition Technology
Editor: Mari F. Burke
ISBN: 979-8-88697-124-8

technology). This is a public version of a sensitive report that GAO issued in April 2021. Information that agencies deemed sensitive has been omitted.

What GAO Recommends

GAO is making two recommendations to each of 13 federal agencies to implement a mechanism to track what non-federal systems are used by employees, and assess the risks of using these systems. Twelve agencies concurred with both recommendations.

U.S. Postal Service concurred with one and partially concurred with the other. GAO continues to believe the recommendation is valid, as described in the report.

What GAO Found

GAO surveyed 42 federal agencies that employ law enforcement officers about their use of facial recognition technology. Twenty reported owning systems with facial recognition technology or using systems owned by other entities, such as other federal, state, local, and non-government entities (see figure).

Agencies reported using the technology to support several activities (e.g., criminal investigations) and in response to COVID-19 (e.g., verify an individual's identity remotely). Six agencies reported using the technology on images of the unrest, riots, or protests following the death of George Floyd in May 2020. Three agencies reported using it on images of the events at the U.S. Capitol on January 6, 2021. Agencies said the searches used images of suspected criminal activity.

All fourteen agencies that reported using the technology to support criminal investigations also reported using systems owned by non-federal entities. However, only one has awareness of what non-federal systems are used by employees. By having a mechanism to track what non-federal systems are used by employees and assessing related risks (e.g., privacy and accuracy-related risks), agencies can better mitigate risks to themselves and the public.

Abbreviations

ABIS	Automated Biometric Identification System
AutoCAT	Automated Credential Authentication Technology
BOP	Federal Bureau of Prisons
CAT-2	Credential Authentication Technology-2
CBP	U.S. Customs and Border Protection
COVID-19	Coronavirus Disease 2019
DHS	Department of Homeland Security
DOD	Department of Defense
FBI	Federal Bureau of Investigation
FDA	Food and Drug Administration
HART	Homeland Advanced Recognition Technology System
ICE	U.S. Immigration and Customs Enforcement
IDENT	Automated Biometric Identification System
NASA	National Aeronautics and Space Administration
OBIM	Office of Biometric Identity Management
OMB	Office of Management and Budget
PFPA	Pentagon Force Protection Agency
Secret Service	U.S. Secret Service
TSA	Transportation Security Administration
VA	U.S. Department of Veterans Affairs

Of all the technologies used to identify people based on their biological and behavioral characteristics, facial recognition most closely mimics how people identify others: by examining their face. Law enforcement can use facial recognition technology to assist criminal investigations, among other activities. For example, the technology can help identify an unknown individual from a photo or image from video surveillance. There are multiple ways to access the technology. Law enforcement may own facial recognition technology, or use technology that is owned by another entity (e.g., federal, state, or non-government entity). However, with use of facial recognition technology expanding, members of Congress and academics have highlighted the importance of understanding what technologies are owned and how they are used by federal law enforcement.

We previously examined aspects of federal agencies' use of facial recognition technology. In September 2020, we reported the U.S. Customs and

Border Protection's (CBP) and Transportation Security Administration's (TSA) use of the technology at U.S. ports of entry.[1] In May 2016, we reported on the Federal Bureau of Investigation's (FBI) use of facial recognition technology.[2]

You asked us to review federal law enforcement use of facial recognition technology. This chapter examines:

1. what federal agencies that employ law enforcement officers own and use facial recognition technology;
2. the type of activities these federal agencies use facial recognition technology to support; and
3. the extent that these federal agencies track employee use of facial recognition technology owned by non-federal entities, including state, local, tribal, territorial, and non-government entities.

This chapter is a public version of a sensitive report that we issued in April 2021.[3] Some federal agencies deemed information in our April report to be sensitive, which must be protected from public disclosure. Therefore, this chapter omits sensitive information about agency ownership and use of facial recognition technology. Although the information provided in this chapter is more limited, the chapter addresses the same objectives as the sensitive report and uses the same methodology.

To address all three objectives, we surveyed 42 federal agencies that employ law enforcement officers. Consistent with our prior work, we define federal law enforcement officers as full-time employees with federal arrest authority and who are authorized to carry firearms while on duty. To identify which agencies employ federal law enforcement officers, we reviewed the Bureau of Justice Statistics' 2016 Census of Federal Law Enforcement

---

[1] GAO, *Facial Recognition: CBP and TSA are Taking Steps to Implement Programs, but CBP Should Address Privacy and System Performance Issues*, GAO-20-568 (Washington, D.C.: September 2, 2020). In this report, GAO made five recommendations to CBP related to its use of facial recognition technology. The Department of Homeland Security concurred with our recommendations, but as of April 2021, has not implemented them.

[2] GAO, *Face Recognition Technology: FBI Should Better Ensure Privacy and Accuracy*, GAO-16-267 (Washington, D.C.: May 16, 2016). In this report, we made six recommendations related to accuracy and privacy regarding the FBI's use of facial recognition technology. The Department of Justice has addressed all six recommendations.

[3] GAO, *Facial Recognition Technology: Federal Law Enforcement Agencies Should Better Assess Privacy and Other Risks*, GAO-21-243SU (Washington, D.C.: April 28, 2021).

Officers.[4] We included 42 of the 86 agencies identified in the 2016 census in our survey population. See Appendix I for a list of the 42 federal agencies we surveyed, and information about why we selected 42 of the 86 agencies in the 2016 census.

To answer our first and second objectives, we administered a survey questionnaire to each of these 42 federal agencies. The questionnaire asked agencies whether at any point from January 2015 through March 2020, they owned a system with facial recognition technology, including systems in the process of being developed.[5] The questionnaire asked agencies that owned a system to complete additional questions, such as the operational status of the system. In addition, we asked agencies whether at any point from April 2018 through March 2020, they used facial recognition technology owned by another entity.[6] We requested additional information, through interviews and written requests, from agencies that reported in their questionnaire that they owned or used facial recognition technology. For example, if an agency reported having a system in operation, we requested privacy impact assessments and system descriptions.

To answer our third objective, we reviewed statutes and regulations, such as the Privacy Act of 1974. In addition, we interviewed or requested information from officials from 14 agencies that reported using (1) non-federal systems, and (2) facial recognition technology to support criminal investigations. We asked these officials about their process for gathering information on what non-federal systems are used by employees, and compared this information against our risk management framework and key

---

[4] Bureau of Justice Statistics, *Federal Law Enforcement Officers, 2016 – Statistical Tables*, NCJ 251922 (Washington, D.C.: October 2019).

[5] We used this time frame because March 2020 was the most recent full month for which information was available when we issued our questionnaire. Also, using a 5-year period allowed us to identify technology that was recently developed but not put in operation, and identify trends in facial recognition search data. In our questionnaire, we stated that the term "own" includes systems that were procured or developed by the respective entity. In addition, we stated that a system with facial recognition technology may include a facial recognition algorithm, hardware, software, and a photo database.

[6] When pretesting our questionnaire, some agencies indicated that they could not guarantee the accuracy of the answers to this question because they did not track the use of systems owned by other entities. In some instances, employees and contractors had to work from their memory on the usage of another entity's systems. To help mitigate this issue, we gathered this information from April 2018 (a 2-year period) instead of January 2015 (a 5-year period). For the purposes of this report, by saying an agency "used" another entity's system, we mean that an agency's offices, employees, and contractors (1) accessed a system owned or operated by another entity, or (2) requested that another entity use its system to conduct a facial recognition search on their behalf.

aspects of *Standards for Internal Control in the Federal Government* (Principles 7, 10, and 16).[7] See Appendix I for additional information on our scope and methodology.

The performance audit upon which this chapter is based was conducted from August 2019 to April 2021 in accordance with generally accepted government auditing standards. Those standards require that we plan and perform the audit to obtain sufficient, appropriate evidence to provide a reasonable basis for our findings and conclusions based on our audit objectives. We believe that the evidence obtained provides a reasonable basis for our findings and conclusions based on our audit objectives. We subsequently worked with the relevant entities from April 2021 to June 2021 to prepare this version of the original sensitive report for public release. This public version was also prepared in accordance with these standards.

Background

How Facial Recognition Technology Works

Biometric technologies can identify individuals by measuring and analyzing biological and behavioral characteristics such as a fingerprint, face, iris, heartbeat, voice, and gait (i.e., a person's manner of walking). Facial recognition is one type of biometric technology. Facial recognition technology uses a photo or still from a video feed of a person—often called a probe or live photo—and converts it into a template, or a mathematical representation of the photo. A matching algorithm can then compare the template to one from another photo and calculate their similarity.[8]

Facial recognition searches generally fall into two categories: verification and identification. Verification (or one-to-one searches) compares a photo to another photo of the same individual. For example, this type of search can help

---

[7] GAO, *Enterprise Risk Management: Selected Agencies' Experiences Illustrate Good Practices in Managing Risk*, GAO-17-63 (Washington, D.C.: Dec. 1, 2016). Also see: GAO, *Standards for Internal Control in the Federal Government*, GAO-14-704G (Washington, D.C.: September 2014). Principle 7 states that management should identify, analyze, and respond to risks related to achieving the defined objectives. Principle 10 states that management should design control activities to achieve objectives and respond to risks. Principle 16 states that management should establish and operate monitoring activities to monitor the internal control system and evaluate the results.

[8] An algorithm is a set of rules that a computer or program follows to compute an outcome. Private companies have developed facial recognition algorithms for a variety of uses.

verify the identity of an individual attempting to unlock a smartphone. Identification (or one-to-many searches) compares a photo from a single individual against a gallery of photos from a number of individuals to determine if there is a potential match. Importantly, identification searches can be used to generate investigative leads (i.e., potential matches) for criminal investigations. Figure 1 shows the process of a facial recognition search, including verification and identification searches.

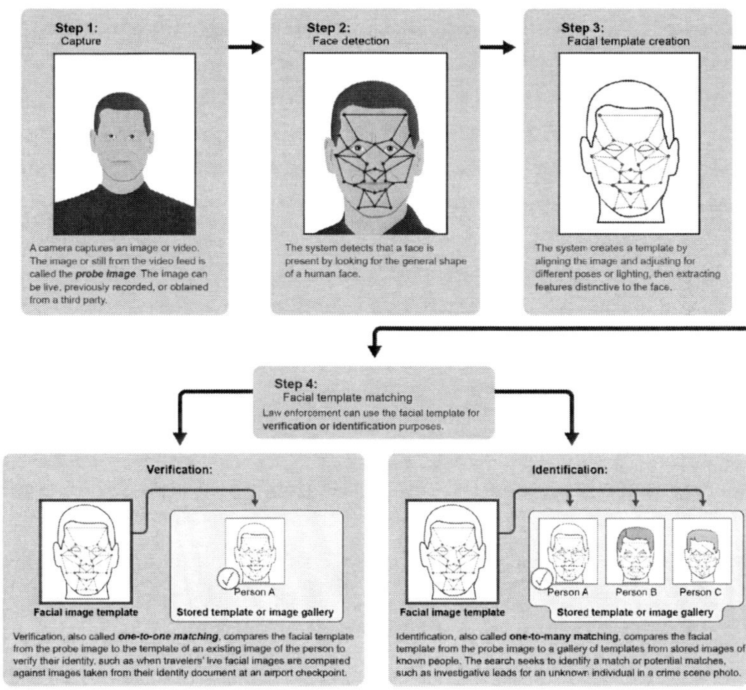

Source: GAO analysis. | GAO-21-518.

Figure 1. Facial Recognition Technology Search Process.

Federal and Non-Federal Systems with Facial Recognition Technology

Federal law enforcement can use systems with facial recognition technology owned by their respective agencies. They can also use systems owned by other government entities, including federal, state, local, tribal, and territorial governments. Moreover, federal law enforcement can use non-government facial recognition service providers, such as Vigilant Solutions and Clearview AI. For example, law enforcement officers with a Clearview AI account can use a computer or smartphone to upload a photo of an unknown individual to Clearview AI's facial recognition system. The system can return search results that show potential photos of the unknown individual, as well as links to the site where the photos were obtained (e.g., Facebook). According to Clearview AI, its system is only used to investigate crimes that have already occurred and not for real-time surveillance. In addition, Clearview AI noted that its system uses images publicly available on the internet, and search results should only be used by law enforcement as investigative leads.

Law enforcement officers may also have access to another entity's system—that is, the officer can log into the system and conduct a facial recognition search. Alternatively, a law enforcement officer can request that another entity use its system to conduct facial recognition searches on their behalf. For example, a federal law enforcement officer may ask a state entity to conduct facial recognition searches on their behalf.

Privacy Laws and Rules

Several statutory requirements govern the protection of personal information by federal agencies, including federal law enforcement's use of facial images. For example, the Privacy Act of 1974 places limitations on agencies' collection, disclosure, and use of personal information maintained in systems of records (e.g., photos). According to Office of Management and Budget (OMB) officials, the Privacy Act and OMB Circular A-130 generally provide that agencies must ensure that privacy requirements apply to systems operated by contractors or other entities on behalf of the Federal Government, which could include facial recognition service providers.

Accuracy of Facial Recognition Technology

The accuracy of facial recognition technology can be characterized in a number of ways. For example, a false positive rate is how often the technology incorrectly declares two images to be a match when they are actually from two different people. In addition, a false negative rate is how often the technology fails to declare two images to be a match when they are actually from the same person. Matching errors can be caused not only by the quality of the facial recognition technology, but also by the quality of the photos used in the matching process and other factors. The National Institute of Standards and Technology has conducted research into the accuracy of facial recognition algorithms. It has evaluated hundreds of commercial facial matching algorithms for accuracy and speed since 2000.

Twenty Federal Agencies Reported Owning or Using Systems with Facial Recognition Technology

Of the 42 agencies we surveyed, 20 reported that they owned a system with facial recognition technology or used another entity's system.[9] As shown in Figure 2, three agencies only owned a system, 12 agencies only used another entity's system, and five agencies both owned a system and used another entity's system. These 20 federal agencies collectively employ roughly 120,000 federal law enforcement officers.[10] According to agencies that owned or used systems, these systems can include hundreds of millions or billions of photos of various types.

Eight Agencies Reported Owning Systems with Facial Recognition Technology

Eight of the 42 federal agencies reported owning 17 systems with facial recognition technology from January 2015 through March 2020.[11] In addition, one of the eight agencies reported that it was in the process of procuring two

---

[9] We asked agencies whether at any point from January 2015 through March 2020, they owned a system with facial recognition technology, including systems in the process of being developed. In addition, we asked agencies whether at any point from April 2018 through March 2020, they used facial recognition technology—that is, their offices, employees, or contractors (1) accessed a system owned/operated by another entity, or (2) requested that another entity use its system to conduct a facial recognition search on their behalf. See the complete list of agencies that received our questionnaire in Appendix I.

[10] These agencies employed roughly 120,000 federal law enforcement officers as of September 30, 2016, based on Bureau of Justice Statistics, *Federal Law Enforcement Officers, 2016 – Statistical Tables*, NCJ 251922 (Washington, D.C.: October 2019).

[11] This report omits some information about systems owned by agencies we surveyed, as the relevant agencies deemed the information sensitive.

systems during this time period, but had not finalized the purchase as of March 2020.[12]

Source: GAO analysis of survey data. | GAO-21-518.

Note: We sent a survey questionnaire to 42 federal agencies that employ law enforcement officers. We asked agencies whether at any point during January 2015 through March 2020, they owned a system with facial recognition technology, including systems in the process of being developed. In addition, we asked agencies whether at any point from April 2018 through March 2020, they used facial recognition technology—that is, their offices, employees, or contractors (1) accessed a system owned or operated by another entity, or (2) requested that another entity use its system to conduct a facial recognition search on their behalf.

The owned system columns include systems in the process of being developed. The National Aeronautics and Space Administration's Office of Protective Services reported that it did not purchase facial recognition technology. However, we included the agency in the owned column because they used a commercial-off-the-shelf product with facial recognition technology to conduct a proof of concept test to determine whether the technology was suitable for its purposes.

Figure 2. Ownership and Use of Facial Recognition Technology Reported by Federal Agencies that Employ Law Enforcement Officers.

[12] The Department of Veterans Affairs Police Service was in the process of procuring the AnyVision and Motorola Avigilon Appearance Search systems as of March 31, 2020. The agency also owned the Aventura and Veritone aiWARE systems during the period of our review.

Table 1 below lists these 19 systems (17 owned and two in procurement) and their statuses as of March 31, 2020. Four of the 19 systems were in operation as of March 31, 2020, and were owned by three agencies: the FBI, Federal Bureau of Prisons, and CBP. Detailed descriptions of the 19 systems can be found in Appendix II.

Detailed System Descriptions, Including Systems Listed in Table 1, Can be Found in Appendix II and III of This Chapter

An example of a system description is illustrated below. These descriptions include how agencies use the systems with facial recognition technology, whether the systems have a photo database, system users, and system statuses, among other information.

DESCRIPTION OF OWNED SYSTEMS

Traveler Verification Service

PURPOSE AND DESCRIPTION

The Traveler Verification Service uses facial recognition technology to verify the identity of international travelers entering and exiting the United States.[1] CBP is testing and deploying the Traveler Verification Service in phases throughout the air, sea, and land travel environments at ports of entry. The system uses real-time capability to compare a traveler's live photo to photos stored in DHS databases,[a] such as passport photos, or to a photo embedded in a travel identification document. Specifically, the system searches DHS databases of photos associated with individuals listed on the travel manifest, and it then creates a prestaged "gallery" of templates created from those photos. CBP uses these "galleries" for matching purposes only and deletes them from the system within 12 hours. CBP plans to use the Traveler Verification Service for all travel environments, but the agency prioritized facial recognition technology in the air environment.

Source: U.S. Customs and Border Protection | GAO-21-518

SYSTEM CHARACTERISTICS

Type of searches:	**One-to-one and One-to-many**[b]
Real-time or near real-time capability:	**Yes**
Includes photo database:	**Yes**
Number of photos in database:	**Not applicable (see description)**

Types of photos in database

- **Passport photos**
- **U.S. entry/exit photos**
- **Visa application photos**

NUMBER OF FACIAL RECOGNITION SEARCHES

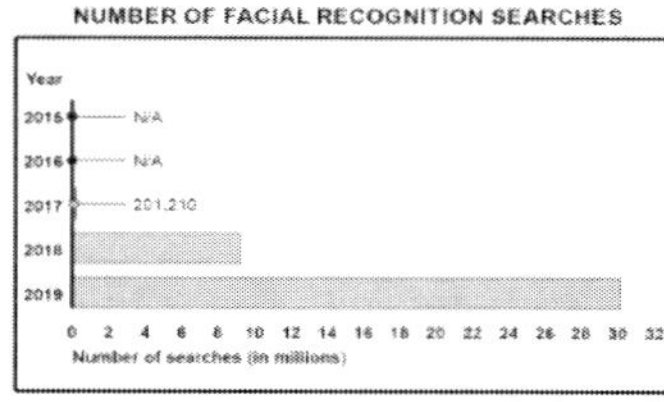

OTHER SYSTEM USERS (EXAMPLES)

Transportation Security Administration

STATUS

As of March 31, 2020, the Traveler Verification Service was in operation and one external entity had access to the system. Specifically, the Transportation Security Administration is piloting the Traveler Verification Service to determine whether it can leverage the capability to improve and automate security processes.

[1]According to CBP, this system is covered by a Privacy Impact Assessment and a System of Records Notice. In addition, while regulations limit CBP's collection of biometric information to certain foreign nationals entering and exiting the United States, CBP's biometric entry-exit capabilities may also capture biometric data (facial images) from exempt foreign nationals and U.S. citizens. However, exempt foreign nationals and U.S. citizens are routinely able to "opt out" of using this technology to verify their identity and can instead choose a manual check of documentation for identity verification.

[a]Although the Traveler Verification Service system accesses photos from other DHS photo databases for matching purposes, it does not store the photos long-term. The photos are deleted from the Traveler Verification Service system within 12 hours. More information on the Traveler Verification Service can be found at GAO-20-568.

[b]By one-to-one, we mean comparing a probe photo (e.g., photo of a person of unknown or unconfirmed identity) to another single photo. By one-to-many, we mean comparing a probe photo to two or more photos.

Page 48 GAO-21-518 Facial Recognition Technology

Source: GAO. | GAO-21-518.

Table 1. Systems with facial recognition technology that federal agencies employing law enforcement officers reported as owned or in procurement, January 2015 through March 2020, and system status

Department	Federal Agency	System Name	System Status as of March 31, 2020[a]
Justice	Federal Bureau of Investigation	Next Generation Identification Interstate Photo System	●
		Horus	○
		Rank One	○
		Automatic Face Detection and Recognition/Cluster Base	⊗
		Camera with Facial Recognition Software	⊗
		NeoFace Reveal	⊗
	Federal Bureau of Prisons	Facial Recognition Access Control System	●
Homeland Security	U.S. Customs and Border Protection	Automated Targeting System	●
		Traveler Verification Service	●
	Transportation Security Administration	Automated Credential Authentication Technology	○
		Credential Authentication Technology-2	○
	U.S. Secret Service	Facial Recognition Pilot	⊗
Veterans Affairs	Police Service	AnyVision	○
		Motorola Avigilon Appearance Search	○
		Aventura	⊗
		Veritone aiWARE	⊗
Defense	Pentagon Force Protection Agency	Briefcam[b]	⊗
		Sirchie	⊗
National Aeronautics and Space Administration	Office of Protective Services	FaceFirst	⊗

Legend: ● In operation; ○ In development (e.g., being tested) or procurement; ⊗ Not in use.

Source: GAO analysis of survey data. I GAO 21 518.

Note: We sent a survey questionnaire to 42 federal agencies that employ law enforcement officers. We asked agencies whether, at any point from January 2015 through March 2020, they owned or were in the process of procuring a system with facial recognition technology, including systems in the process of being developed. This table omits some information about systems owned by agencies we surveyed, as the relevant agency deemed the information sensitive.

[a]The system status category "Not in use" refers to systems that agencies owned from January 1, 2015 through March 31, 2020; however, the system was not in use as of March 31, 2020. For example, agencies reported that they no longer used a system, or tested a system and determined it was not suitable for their purposes.

[b]The Briefcam system was in operation as of March 31, 2020; however, according to Pentagon Force Protection Agency officials, the facial recognition technology component of this system was not in operation. As such, the status of this system is marked "Not in use."

Seventeen Federal Agencies Reported Using Systems Owned by Other Entities

Seventeen of the 42 federal agencies reported using another entity's system with facial recognition technology from April 2018 through March 2020. Of the 17 agencies, 15 reported using systems owned by another federal entity; 14 reported using systems owned by state, local, tribal, or territorial entities; and 11 reported using systems owned by non-government entities. Furthermore, nine of the 17 agencies reported using systems owned by all three types of entities. See Table 2 for additional information.

As discussed above, 15 agencies reported using a system owned by another federal entity. Ten of these agencies reported using systems owned by other federal agencies we surveyed. For example, the Transportation Security Administration reported using the U.S. Customs and Border Protection's Traveler Verification Service. Agencies also reported using systems that were owned by federal entities that we did not survey. Specifically, based on survey responses, multiple agencies reported using the three systems listed in the table below. More information on each of these systems can be found in Appendix III.

Fourteen federal agencies reported using systems owned by state, local, tribal, and territorial entities. For example, FBI's Facial Analysis, Comparison, and Evaluation Services had memorandums of understanding with certain state agencies, allowing it to leverage the state-owned systems for facial recognition searches.[13] According to the FBI, these state-owned systems include driver's license photos, mugshots, or corrections photos.

Eleven agencies we surveyed used systems with facial recognition technology belonging to non-government entities, including Clearview AI (10 agencies) and Vigilant Solutions (five agencies).[14] Ten agencies reported having used non-government facial recognition service providers on a free trial basis. For example, according to the U.S. Postal Inspection Service, it had a free trial with Vigilant Solutions that lasted approximately 10 months in 2017.

---

[13] The FBI's Facial Analysis, Comparison, and Evaluation Services is located in the Investigative Services Support Unit of the Criminal Justice Information Services Division. It conducts facial recognition searches on the FBI's Next Generation Identification Interstate Photo System and can request external partners perform searches on their facial recognition systems to support FBI active investigations.

[14] Information regarding the extent that agencies used Clearview AI and Vigilant Solutions has been omitted from this report, as some agencies deemed the information sensitive.

Table 2. Reported use of other entities' facial recognition technology by federal agencies that employ law enforcement officers

Type of Entity That Owned System					
			Non-Government[a]		
Federal Agency That Used System	Other Federal	State, Local, Tribal, Territorial	Clearview AI	Vigilant Solutions	Other Non-Government[b]
Bureau of Diplomatic Security	✓	✓	✓	✓	✓
U.S. Customs and Border Protection	✓	✓	✓	✓	✓
U.S. Marshals Service	✓	✓	✓	✓	✓
Bureau of Alcohol, Tobacco, Firearms and Explosives	✓	✓	✓	✓	—
U.S. Immigration and Customs Enforcement	✓	✓	✓	—	✓
U.S. Postal Inspection Service	✓	✓	✓	✓	—
Drug Enforcement Administration	✓	✓	✓	—	—
Federal Bureau of Investigation	✓	✓	✓	—	—
U.S. Secret Service	✓	✓	✓	—	—
U.S. Capitol Police	✓	✓	—	—	—
U.S. Fish and Wildlife Service	✓	✓	—	—	—
Food and Drug Administration, Office of Criminal Investigations	✓	✓	—	—	—
Internal Revenue Service, Criminal Investigation Division	✓	✓	—	—	—
U.S. Park Police	—	✓	✓	—	—
Administrative Office of the U.S. Courts, U.S. Probation and Pretrial Services	—	—	—	—	✓
Pentagon Force Protection Agency	✓	—	—	—	—
Transportation Security Administration	✓	—	—	—	—
Total	15	14	10	5	5

Legend:

✓Agency used a system owned by the respective entity (or entity type) at any point from April 2018 through March 2020. For federal, state, local, tribal, and territorial entities, the term "used" includes an agency's offices, employees, or contractors (1) accessing a system owned/operated by the respective entity type, or (2) requesting that the respective entity type use its system to conduct a facial recognition search on the agency's behalf. For non-government entities, the term "used" means the agency's offices, employees, or contractors submitted photos to the respective non-government service provider for the purpose of conducting a facial recognition search.

— Agency did not use a system owned by the respective entity (or entity type) at any point from April 2018 through March 2020.

Source: GAO analysis of survey data. I GAO-21-518.

[a]Some agencies reported that they only used Clearview AI or Vigilant Solutions on a free trial basis, and thus, did not enter into a formal contract with the service provider.

[b]Other non-government entities that agencies reported using included Amazon Rekognition, BI SmartLink, and Giant Oak Social Technology, among others.

Table 3. Select systems with facial recognition technology owned by federal agencies

System Name	Federal Agency That Owns System
Automated Biometric Identification System (ABIS)	Defense Forensic Science Center (Department of Defense)
Automated Biometric Identification System (IDENT)	Office of Biometric and Identity Management (Department of Homeland Security)
Integrated Biometric System	Bureau of Consular Affairs (Department of State)

Source: GAO analysis of survey data. I GAO-21-518.

Examples of Federal Agencies Partnering with State and Local Entities

- Food and Drug Administration's (FDA) Office of Criminal Investigations reported using the Georgia Department of Driver Services' facial recognition technology. FDA reported using the technology to verify the identity of an individual under investigation who had assumed a stolen identity.
- U.S. Immigration and Customs Enforcement (ICE) reported helping to fund the development of a system with facial recognition technology that Lehigh County Regional Investigation and Intelligence Center (the Center) will own. Specifically, ICE and the Center are developing the National Capital Region Gang Intelligence Application to combat transnational gangs, according to ICE officials. Officials told us that once developed, ICE can use the system's facial recognition technology to compare images of unknown individuals to a gallery of known and suspected gang members.
- U.S. Park Police reported that it asked a state agency to use facial recognition technology to help identify a deceased individual in a park. The state agency used a free trial with a non-government service provider to conduct the search, according to U.S. Park Police.

Source: GAO analysis of survey data, zfmbek/stock.adobe.com. I GAO-21-518.

Agencies Reported Using Systems that Varied in the Number and Type of Photos

Types of Photos Used by Federal Agencies that Employ Law Enforcement Officers

Federal agencies reported using a number of systems with facial recognition technology. The following list includes examples of the types of photos included in these systems, as reported by system owners and users:

- Mug shot
- Publicly available on the internet
- Passport
- Visa application
- U.S. entry/exit
- Video/Closed Circuit Television
- Terrorist Screening Database
- Foreign nationals and U.S. citizens who are known or suspected threats to the nation
- Employee
- State identification
- Driver's license
- Corrections
- Individuals under supervision

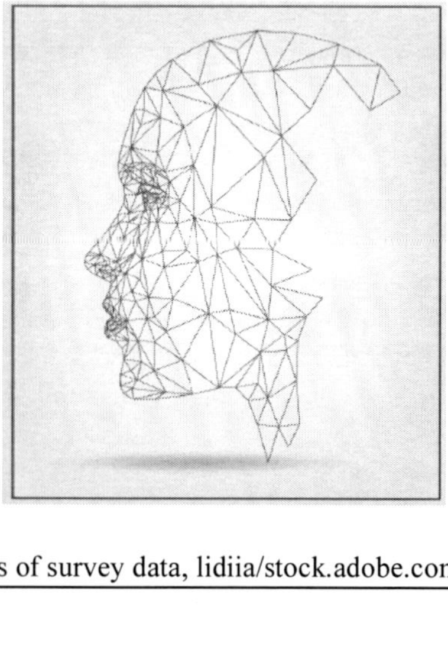

Source: GAO analysis of survey data, lidiia/stock.adobe.com. | GAO-21-518.

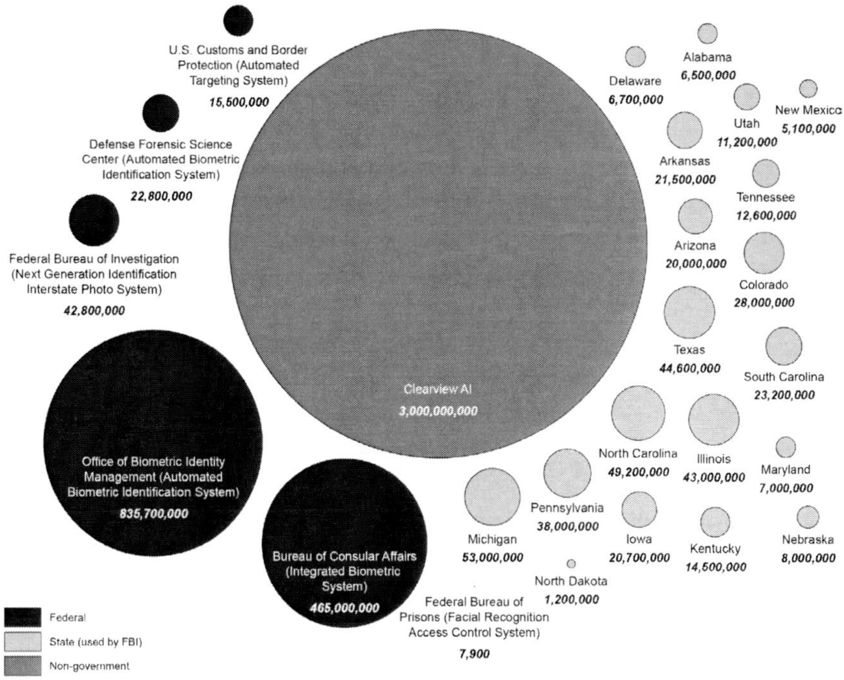

Source: GAO analysis of information provided by system users or owners. | GAO-21-518.

Note: We sent a survey questionnaire to 42 federal agencies that employ law enforcement officers. This figure includes examples of systems used by one or more federal agencies we surveyed. It does not include all systems used by these agencies. The figure includes the number of photos stored in the respective entity's system with facial recognition technology, as of March 31, 2020.

The same individual may be included in multiple photos within one photo database or across multiple databases, and the same photo can exist within multiple databases. Some entities providing these numbers indicated they were estimates. The number of photos for federal and non-government entities were reported by the respective system owner. The number of photos for state entities were all reported by the Federal Bureau of Investigation (FBI). Specifically, the FBI's Facial Analysis, Comparison, and Evaluation Services has memorandums of understanding with several state agencies, allowing it to leverage the state-owned systems for facial recognition searches. The FBI provided the number of photos they can access via these memorandums of understanding.

Figure 3. Selected Federal, State, and Non-government Systems with Facial Recognition Technology Used by Federal Agencies that Employ Law Enforcement Officers, and the Number of Photos in Them.

Federal agencies reported using numerous systems with facial recognition technology, and sometimes these systems included stored photos. The number and types of photos within these systems can vary, based on information reported by agencies and system owners. For example, as of March 31, 2020, the Bureau of Prisons' Facial Recognition Access Control System included roughly 8,000 photos of its employees and contractors. Other systems include millions or billions of photos. For example, the Office of Biometric and Identity Management reported that its Automated Biometric Identification System (IDENT) included roughly 836 million facial images as of March 31, 2020. According to agency officials, the types of photos include visa application, passport, mug shot, and others. Clearview AI told us its system includes roughly 3 billion publicly available photos gathered from the internet.

Figure 3 above shows examples of federal, state, and non-government systems with facial recognition technology that federal agencies reported using, and the number of photos in them. Appendix II includes additional information on photos, including the number and type of photos in systems owned by federal agencies we surveyed.

Federal Agencies Reported Using Systems with Facial Recognition Technology to Support Various Activities

Most federal agencies that owned or used facial recognition technology reported using it to support criminal investigations. Agencies also reported using facial recognition technology for activities such as surveillance, traveler verification, and research and education.

Of the 20 federal agencies that owned or used facial recognition technology, 14 reported using the technology to support criminal investigations. For example, the FBI's Next Generation Identification Interstate Photo System allows users to search a database of over 40 million photos. The system returns a list of potential candidates that law enforcement can use to generate investigative leads. According to the FBI, the system has been used for investigations of violent crimes, credit card and identity fraud, missing persons, and bank robberies, among others. The Department of Homeland Security's Office of Biometric Identity Management offers a similar service to its partners (e.g., U.S. Immigration and Customs Enforcement). Specifically, the agency's Automated Biometric Identification System can be used to search a photo of an unknown individual and provide potential matches (i.e., generate leads) to support criminal investigations. Federal agencies also reported using state, local, and non-government systems to support criminal investigations.

Table 4. Federal agency reported use of facial recognition technology on images of individuals suspected of violating the law during civil unrest, riots, or protests, May through August 2020

Federal Agency	How Agency Reported Using Facial Recognition Technology
Bureau of Alcohol, Tobacco, Firearms and Explosives	In a single instance, used facial recognition technology owned by another law enforcement entity. The search was conducted to help identify an individual suspected of violating the law during the period of civil unrest, riots, or protests.
U.S. Capitol Police	Requested that the Montgomery County Department of Police (Montgomery County, Maryland) conduct facial recognition searches to assist with a criminal investigation. The purpose of the searches was to help identify individuals that confronted and made threats to a member of Congress and the member's spouse outside the White House during the period of civil unrest, riots, or protests.
Federal Bureau of Investigation	Created a digital media tip line and solicited images of people involved in criminal activity during the period of civil unrest, riots, or protests. The agency sought to identify or locate criminal suspects seen in images and video depicting criminal behavior by conducting facial recognition searches using its Next Generation Identification Interstate Photo System.
U.S. Marshals Service	Used a non-government facial recognition service provider, to conduct facial recognition searches related to criminal investigations on images from the period of civil unrest, riots, or protests.
U.S. Park Police	Requested that the Maryland National Capital Park Police conduct a facial recognition search using an image from Twitter to identify an individual who allegedly assaulted an officer during the period of civil unrest, riots, or protests. The search was conducted on the National Capital Region Facial Recognition Investigative Leads System. The subject was ultimately charged with Felony Civil Disorder and two counts of Assault on a Police Officer.
U.S. Postal Inspection Service	Used Clearview AI to help identify individuals suspected of criminal activity that took place in conjunction with the period of civil unrest, riots, or protests. This criminal activity included damaging U.S. Postal Service property, stealing mail, opening mail, burglarizing U.S. Postal Service buildings, and committing arson.

Source: GAO analysis of survey data. | GAO-21-518.

Six agencies reported using facial recognition technology during May through August 2020 to support criminal investigations related to civil unrest, riots, or protests.[15] Following the death of George Floyd while in the custody of the Minneapolis, Minnesota police department on May 25, 2020, nationwide civil unrest, riots, and protests occurred.[16] Six agencies told us that they used images from these events to conduct facial recognition searches during May through August 2020 in order to assist with criminal investigations (see Table 4). All six agencies reported that these searches were on images of individuals suspected of violating the law.

In addition, with regard to the January 6, 2021 events at the U.S. Capitol complex, three agencies reported using facial recognition technology to support criminal investigations related to the civil unrest, riots, or protests.[17] The three agencies reported using the technology to support criminal investigations as follows:

- U.S. Capitol Police used Clearview AI to help generate investigative leads. The agency also requested that another federal agency use its system to conduct facial recognition searches on behalf of the U.S. Capitol Police.
- CBP used its Automated Targeting System to conduct searches at the request of another federal agency.
- Bureau of Diplomatic Security used the Department of State's Integrated Biometric System to conduct searches at the request of another federal agency.

Agencies also reported using facial recognition technology beyond criminal investigations. The following list includes examples of use cases, as reported by agencies. Appendix II includes more information on these use cases and others.

---

[15] We requested this information from 17 agencies that indicated in their questionnaire response as (1) having a system with facial recognition technology that was in operation, or (2) using another entity's system. See more information on our methodology in appendix I.

[16] In September 2020, we reported that federal agencies deployed 16 tactical teams in relation to the civil unrest and protests. See: GAO, *Federal Tactical Teams: Characteristics, Training, Deployments, and Inventory*, GAO-20-710 (Washington, D.C.: September 2020).

[17] We requested this information from 17 agencies that indicated in their questionnaire response as (1) having a system with facial recognition technology that was in operation, or (2) using another entity's system. See more information on our methodology in appendix I. Twelve agencies reported that they did not use the technology for these purposes, three agencies reported using the technology, and two agencies told us they could not answer our questions because the information pertains to ongoing investigations.

- *Surveillance.* The U.S. Secret Service (Secret Service) piloted a system with facial recognition technology to determine whether it could be incorporated into the agency's White House Complex security operations. Specifically, the Secret Service stored photos of 23 volunteer employees within the system. As volunteers moved throughout the White House Complex, their images were captured by closed-circuit television cameras. In real time, the system compared the stored photos to images from the video footage to determine whether they represented the same individual. Secret Service told us it did not plan to implement the system based on the results of the pilot.
- *Response to Coronavirus Disease 2019 (COVID-19).* In response to COVID-19, the Administrative Office of the U.S. Courts, Probation and Pretrial Services office began using facial recognition technology. The technology allowed individuals under court-ordered supervision to verify their identity via a smart phone application rather than physical contact with a probation or pretrial officer. According to agency officials, the program is limited to voluntary use by individuals under supervision in connection with court-ordered location or alcohol monitoring.
- *Traveler verification.* CBP's Traveler Verification Service uses facial recognition technology to verify the identity of travelers entering and exiting the United States.[18] CBP is testing and deploying the Traveler Verification Service in phases throughout the air, sea, and land travel environments at ports of entry. The system uses real-time capability to compare a traveler's live photo to photos stored in Department of Homeland Security databases, such as passport photos, or to a photo embedded in a travel identification document.
- *Area access.* The Federal Bureau of Prisons (BOP) uses its Facial Recognition Access Control System to authenticate entry into secure network operations centers at certain BOP facilities. The system verifies BOP employees' identities using facial recognition

---

[18] While regulations limit CBP's collection of biometric information to certain foreign nationals entering and exiting the United States, CBP's biometric entry-exit capabilities may also capture biometric data (facial images) from exempt foreign nationals and U.S. citizens. However, exempt foreign nationals and U.S. citizens are routinely able to "opt out" of using this technology to verify their identity and can instead choose a manual check of documentation for identity verification. For more information, see GAO-20-568.

technology, and once confirmed, employees can enter the operations centers.

- *Research and education.* The FBI is using systems for research and education purposes. For example, it is examining how well systems perform when used on its casework. In addition, the bureau is trying to determine whether these systems could be incorporated into its one-to-one comparisons process. Currently, FBI forensic examiners manually compare two images to validate whether faces within the images represent the same individual (i.e., one-to-one comparisons). The FBI is researching whether it would be beneficial to use a facial recognition system in addition to forensic examiners. It is also using systems in educational settings to demonstrate how facial recognition technology works.

Most Agencies Do Not Track Non-Federal Systems in Use or Related Risks

Thirteen federal agencies do not have awareness of what non-federal systems with facial recognition technology are used by employees.[19] These agencies have therefore not fully assessed the potential risks of using these systems, such as risks related to privacy and accuracy. Most federal agencies that reported using non-federal systems did not own systems. Thus, employees were relying on systems owned by other entities, including non-federal entities, to support their operations.

Tracking Use of Non-Federal Systems

We found that 13 of 14 agencies that reported using non-federal systems do not have complete, up-to-date information on what non-federal systems are used by employees.[20] For example, when we requested information from one

---

[19] Throughout this section, when we say that an agency did not have awareness of what systems are used by employees, we are referring to both the agency's employees and contractors. By non-federal systems, we are referring to systems owned by state, local, tribal, territorial, and non-government entities.

[20] By complete, up-to-date information, we mean that an agency has ongoing knowledge of what non-federal systems with facial recognition technology are used by employees. Fifteen agencies reported using non-federal systems; however, we excluded U.S. Probation and

of the agencies about its use of non-federal systems, agency officials told us they had to poll field division personnel because the information was not maintained by the agency. These agency officials also told us that the field division personnel had to work from their memory about their past use of non-federal systems, and that they could not ensure we were provided comprehensive information about the agency's use of non-federal systems. Officials from another agency initially told us that its employees did not use non-federal systems; however, after conducting a poll, the agency learned that its employees had used a non-federal system to conduct more than 1,000 facial recognition searches.

One agency—the U.S. Immigration and Customs Enforcement—reported that it was in the process of implementing a mechanism to track what non-federal systems are used by employees.[21] According to U.S. Immigration and Customs Enforcement officials, in November 2020 they were in the process of developing a list of approved facial recognition technologies that employees can use. In addition, log-in sheets will be made available to employees, allowing supervisors to monitor employee use of the technologies. The agency will allow the use of non-reviewed systems under exigent circumstances; however, supervisor approval is required. The use of systems that have been reviewed and not approved for use is strictly prohibited, even in exigent circumstances, according to U.S. Immigration and Customs Enforcement.

However, the other 13 agencies do not have complete, up-to-date information because they do not regularly track this information and have no mechanism in place to do so (see Table 5).[22] For example, the Criminal Investigation Division within the Internal Revenue Service told us it does not track what non-federal systems are used by employees because it is not the owner of these technologies. Similarly, the U.S. Postal Inspection Service said it did not track what systems employees use because it is the responsibility of the system owner to govern use of the system.

Pretrial Services because it does not use facial recognition technology to support criminal investigations. All 14 agencies discussed in this section reported using the technology to support criminal investigations.

[21] According to U.S. Immigration and Customs Enforcement officials, only employees within its Homeland Security Investigations will be subject to the procedures, as only employees within this component of the agency use facial recognition technology.

[22] We asked agencies whether they had a mechanism to track *what systems* were used by employees, not whether agencies track *each individual use of a system* by employees.

Table 5. Federal agency tracking of employee use of non-federal systems with facial recognition technology

Federal Agency	Have Mechanism to Track What Non-Federal Systems Are Used by Employees
U.S. Immigration and Customs Enforcement	Yes
Bureau of Alcohol, Tobacco, Firearms and Explosives	No
Bureau of Diplomatic Security	No
U.S. Capitol Police	No
U.S. Customs and Border Protection	No
Drug Enforcement Administration	No
Federal Bureau of Investigation	No
U.S. Fish and Wildlife Service	No
Food and Drug Administration, Office of Criminal Investigations	No
Internal Revenue Service, Criminal Investigation Division	No
U.S. Marshals Service	No
U.S. Park Police	No
U.S. Postal Inspection Service	No
U.S. Secret Service	No

Source: GAO analysis of agency information. I GAO-21-518.

Note: Federal agencies marked "No" may have known that employees used certain systems, but they do not have a mechanism to provide complete, up-to-date information of what systems are used by employees.

Standards for Internal Control in the Federal Government state that agencies should design and implement controls to help achieve agencies' objectives, which, in this case, is to conduct investigative activities.[23] These standards also state that ongoing monitoring—such as regular oversight that would provide visibility into the non-federal systems used by employees—should be performed continually in the course of normal operations.

Internal control standards further underscore that management should identify, analyze, and respond to risks related to achieving the defined objectives. Additionally, we have previously reported that enterprise risk management can help federal agencies assess risks, such as those related to the use of non-federal systems.[24] For example, assembling a list of risks can help an agency identify threats that could limit its ability to achieve goals and objectives.

[23] GAO-14-704G.

[24] GAO-17-63.

By implementing a mechanism to track what non-federal systems with facial recognition technology are used by employees, agencies will be able to have visibility into the systems that employees are relying on to assist with investigative activities.[25] Gathering this information on a continuous basis can serve as an important initial step to identifying risks associated with non-federal systems. When asked about the potential implementation of a mechanism to track what systems are used by employees, seven of the 13 agencies told us that it was feasible or did not express any specific concerns. The other six agencies did not comment.

Assessing Risks Related to Non-Federal Systems

As 13 federal agencies do not have awareness of non-federal systems used by employees, they cannot fully assess the risks of using these systems. Numerous risks to federal agencies and the public can accompany the use of facial recognition technology. In particular, these risks can relate to privacy and the accuracy of a system.

Several privacy-related requirements govern the protection of personal information by federal agencies, including federal law enforcement's use of facial images. For example, the Privacy Act of 1974 places limitations on agencies' collection, disclosure, and use of personal information maintained in systems of records (e.g.. photos).[26] Additionally, according to OMB officials, the E-Government Act of 2002 and OMB Privacy policy requirements necessitate that when an agency procures information technology that processes personally identifiable information from outside services, such as a third-party vendor or state or local government, agencies must conduct an assessment of the privacy implications.[27]

When agencies use facial recognition technology without first assessing the privacy implications and applicability of privacy requirements, there is a

---

[25] This sentence describes the potential benefits of tracking what systems are used by employees, and not each individual use of a system by employees.

[26] See 5 U.S.C. § 552a(4). Per OMB guidance and the Federal Acquisition Regulation, agency obligations to maintain privacy protections and adhere to the Privacy Act of 1974 obligations extend to information technology systems that are used or operated by contractors or other entities on behalf of the federal government or that collect or maintain federal information on behalf of the federal government. See Office of Management and Budget Memorandum (OMB) Circular A-130, Managing Information as a Strategic Resource (July 28, 2016); Federal Acquisition Regulation, Subpart 24.1 Protection of Individual Privacy.

[27] See Pub. L. No. 107-347, 116 Stat. 2899 (2002).

risk that they will not adhere to privacy-related laws, regulations, and policies. There is also a risk that non-federal system owners will share sensitive information (e.g., photo of a suspect) about an ongoing investigation with the public or others. In addition, privacy advocacy organizations, government agencies, academics, and some industry representatives have raised privacy and security concerns. For example, there is a risk that data sets with personal information could be subject to breaches, resulting in sensitive biometric data being revealed to unauthorized entities. Because a person's face is distinctive, permanent, and therefore irrevocable, a breach involving data derived from a face may have more serious consequences than the breach of other information, such as passwords, which can be changed.

The U.S. Immigration and Customs Enforcement assessed privacy risks associated with its use of facial recognition technology, including non-federal systems.[28] The assessment was reviewed by the Chief Privacy Officer for the Department of Homeland Security. In the assessment, the agency identified privacy risks and what, if any, actions it can take to mitigate these risks. For example, as part of its process, before using a non-federal systems the agency plans to ensure that:

- methods of transmission of a probe photo are properly encrypted,
- no probe photos submitted by U.S. Immigration and Customs Enforcement are retained by the system owner or shared with other parties, and
- appropriate safeguards for housing sensitive personally identifiable information exist.

Accuracy rates can help agencies determine how often facial recognition technology correctly or incorrectly declares that two or more images match. Matching errors can be caused by the quality of the facial recognition technology, and other factors such as the quality of photos used in the matching process. Although the accuracy of facial recognition technology has increased dramatically in recent years, risks still exist that searches will provide inaccurate results. For example, if a system is not sufficiently accurate, it could unnecessarily identify innocent people as investigative leads. The system could also miss investigative leads that could otherwise have been revealed. In December 2019, the National Institute of Standards and

[28] Department of Homeland Security, *Privacy Impact Assessment for the ICE Use of Facial Recognition Services*, DHS/ICE/PIA-054 (May 13, 2020).

Technology reported that facial recognition algorithms it tested differed in accuracy widely by race, ethnicity, or country of origin, as well as by gender and age.[29] In addition, some members of Congress, privacy groups, and others have expressed concerns that facial recognition technology's higher error rates for certain demographics could result in disparate treatment, profiling, or other adverse consequences for members of these populations.

The U.S. Immigration and Customs Enforcement, as part of the risk assessment described earlier, also considered the accuracy of non-federal systems. According to the assessment, the agency will leverage resources such as the National Institute of Standards and Technology's testing on the accuracy and bias of systems. Additionally, the agency said it will conduct non-scientific tests to gain insight into the veracity of a system.

However, as the other 13 federal agencies do not have awareness of non-federal systems used by employees, they cannot fully assess the risks (e.g., privacy and accuracy-related risks) of using these systems. As described earlier, we have previously reported that assembling a list of risks can help an agency identify threats to achieving its goals and objectives. In addition, prioritizing identified risks and selecting the most appropriate treatment strategy to manage the risks are important next steps.

Without assessing the risks of using non-federal systems, agencies are more susceptible to risks that negatively affect their ability to meet stated goals and objectives. For example, agencies cannot ensure appropriate privacy safeguards are in place to prevent the mishandling of personal information when using non-federal systems. In addition, agencies cannot ensure they are using systems sufficiently accurate for their purposes.[30] Failure to appropriately assess risks when using non-federal systems with facial recognition technology could ultimately result in a reputational catastrophe, hindering an agency's efforts to meet its core mission for years.

---

[29] National Institute of Standards and Technology, *Face Recognition Vendor Test (FRVT) Part 3: Demographic Effects,* NIST Interagency or Internal Report 8280 (Dec. 19, 2019). The National Institute of Standards and Technology reported that it tested 189 mostly commercial algorithms from 99 developers and that performance differences varied by the algorithms tested, with some performing better than others. For a small number of the one-to-many algorithms, differences in false positives across demographic groups were undetectable. The extent of performance differences varied by the developer, type of error, and quality of the facial images.

[30] This sentence describes the potential benefits of understanding and managing accuracy-related risks, which may or may not result in agencies testing the accuracy of individual systems used by employees.

Conclusion

Facial recognition technology is a powerful tool used by the federal law enforcement community. Federal agencies that employ law enforcement officers rely on systems with facial recognition technology, and the potentially millions or billions of photos stored in these systems, to help generate investigative leads and solve crimes. However, 13 federal agencies cannot assess the risks of using non-federal systems because they are unaware of what systems are used by employees. By implementing a mechanism to track what non-federal systems are used by employees, agencies will have better visibility into the technologies they rely upon to conduct criminal investigations. In addition, by assessing the risks of using these systems, including privacy and accuracy-related risks, agencies will be better positioned to mitigate any risks to themselves and the public.

Recommendations for Executive Action

We are making the following 26 recommendations:

The Director of the Bureau of Alcohol, Tobacco, Firearms and Explosives should implement a mechanism to track what non-federal systems with facial recognition technology are used by employees to support investigative activities. (Recommendation 1)

The Director of the Bureau of Alcohol, Tobacco, Firearms and Explosives should, after implementing a mechanism to track non-federal systems, assess the risks of using such systems, including privacy and accuracy-related risks. (Recommendation 2)

The Administrator for the Drug Enforcement Administration should implement a mechanism to track what non-federal systems with facial recognition technology are used by employees to support investigative activities. (Recommendation 3)

The Administrator for the Drug Enforcement Administration should, after implementing a mechanism to track non-federal systems, assess the risks of using such systems, including privacy and accuracy-related risks. (Recommendation 4)

The Director of the FBI should implement a mechanism to track what non-federal systems with facial recognition technology are used by employees to support investigative activities. (Recommendation 5)

The Director of the FBI should, after implementing a mechanism to track non-federal systems, assess the risks of using such systems, including privacy and accuracy-related risks. (Recommendation 6)

The Director of the U.S. Marshals Service should implement a mechanism to track what non-federal systems with facial recognition technology are used by employees to support investigative activities. (Recommendation 7)

The Director of the U.S. Marshals Service should, after implementing a mechanism to track non-federal systems, assess the risks of using such systems, including privacy and accuracy-related risks. (Recommendation 8)

The Commissioner of CBP should implement a mechanism to track what non-federal systems with facial recognition technology are used by employees to support investigative activities. (Recommendation 9)

The Commissioner of CBP should, after implementing a mechanism to track non-federal systems, assess the risks of using such systems, including privacy and accuracy-related risks. (Recommendation 10)

The Director of the Secret Service should implement a mechanism to track what non-federal systems with facial recognition technology are used by employees to support investigative activities. (Recommendation 11)

The Director of the Secret Service should, after implementing a mechanism to track non-federal systems, assess the risks of using such systems, including privacy and accuracy-related risks. (Recommendation 12)

The Director of the U.S. Fish and Wildlife Service should implement a mechanism to track what non-federal systems with facial recognition technology are used by employees to support investigative activities. (Recommendation 13)

The Director of the U.S. Fish and Wildlife Service should, after implementing a mechanism to track non-federal systems, assess the risks of using such systems, including privacy and accuracy-related risks. (Recommendation 14)

The Chief of the U.S. Park Police should implement a mechanism to track what non-federal systems with facial recognition technology are used by employees to support investigative activities. (Recommendation 15)

The Chief of the U.S. Park Police should, after implementing a mechanism to track non-federal systems, assess the risks of using such systems, including privacy and accuracy-related risks. (Recommendation 16)

The Assistant Secretary of the Bureau of Diplomatic Security should implement a mechanism to track what non-federal systems with facial recognition technology are used by employees to support investigative activities. (Recommendation 17)

The Assistant Secretary of the Bureau of Diplomatic Security should, after implementing a mechanism to track non-federal systems, assess the risks of using such systems, including privacy and accuracy-related risks. (Recommendation 18)

The Assistant Commissioner of the Food and Drug Administration's Office of Criminal Investigations should implement a mechanism to track what non-federal systems with facial recognition technology are used by employees to support investigative activities. (Recommendation 19)

The Assistant Commissioner of the Food and Drug Administration's Office of Criminal Investigations should, after implementing a mechanism to track non-federal systems, assess the risks of using such systems, including privacy and accuracy-related risks. (Recommendation 20)

The Chief of the Internal Revenue Service's Criminal Investigation Division should implement a mechanism to track what non-federal systems with facial recognition technology are used by employees to support investigative activities. (Recommendation 21)

The Chief of the Internal Revenue Service's Criminal Investigation Division should, after implementing a mechanism to track non-federal systems, assess the risks of using such systems, including privacy and accuracy-related risks. (Recommendation 22)

The Chief Postal Inspector of the U.S. Postal Inspection Service should implement a mechanism to track what non-federal systems with facial recognition technology are used by employees to support investigative activities. (Recommendation 23)

The Chief Postal Inspector of the U.S. Postal Inspection Service should, after implementing a mechanism to track non-federal systems, assess the risks of using such systems, including privacy and accuracy-related risks. (Recommendation 24)

The Chief of Police, U.S. Capitol Police, should implement a mechanism to track what non-federal systems with facial recognition technology are used by employees to support investigative activities. (Recommendation 25)

The Chief of Police, U.S. Capitol Police, should, after implementing a mechanism to track non-federal systems, assess the risks of using such systems, including privacy and accuracy-related risks. (Recommendation 26)

Agency Comments and Our Evaluation

We provided a draft of this product for comment to the 21 federal departments and other entities (i.e., entities) that responded to our survey.[31] We made recommendations to eight of the 21 entities, and these eight entities generally concurred with our recommendations.[32]

Seven of the eight entities provided written comments, which are reproduced in appendices IV through X and summarized below.[33]

- U.S. Department of Health and Human Services concurred with our recommendations.
- Department of Homeland Security concurred with our recommendations and provided technical comments, which we incorporated as appropriate.
- Department of the Interior concurred with our recommendations.
- Department of Justice's Federal Bureau of Investigation concurred with our recommendations and provided technical comments, which we incorporated as appropriate.
- Department of State concurred with our recommendations and provided technical comments, which we incorporated as appropriate.
- Department of the Treasury concurred with our recommendations, stating that the Internal Revenue Service's Criminal Investigation Division has mechanisms in place capable of tracking non-federal systems used by employees. However, during our review, the Criminal Investigation Division told us it does not track what systems are used by employees because it is not the owner of these technologies. Moreover, as noted in our report, the Criminal Investigation Division used facial recognition technology owned by other governmental entities at the federal and state, local, tribal, or

---

[31] Although we surveyed 42 federal agencies, we generally provide draft products for comment to the respective department (e.g., Department of Homeland Security) rather than the individual agencies (e.g., U.S. Customs and Border Protection) within a department. See the list of 21 departments and other entities (e.g., Amtrak) that received the draft product at Table 6 in Appendix I.

[32] We made recommendations to 13 of the 42 federal agencies that we surveyed. These 13 agencies are located within eight federal departments and entities (i.e., entities), and as discussed, these eight entities generally concurred with our recommendations.

[33] As discussed earlier, this report is a public version of a sensitive report that we issued in April 2021. Five of the seven entities asked that we reprint their comments from the sensitive report, which are reproduced in appendices VI through X.

territorial level. We therefore continue to believe that the Criminal Investigation Division should, as part of implementing these recommendations, ensure mechanisms are in place to appropriately track what non-federal systems (e.g., systems owned by other federal and state agencies) are used to support its investigative activities.

- U.S. Postal Service concurred, in part, with our recommendation to develop a mechanism to track what non-federal systems employees use. Specifically, the agency told us that the U.S. Postal Inspection Service currently tracks employee use of certain non-federal systems. Thus, the agency said it only needs to develop a mechanism to track other non-federal systems that employees use. Our report acknowledges that agencies may have had awareness of certain non-federal systems used by employees. However, U.S. Postal Inspection Service does not have a mechanism to provide complete, up-to-date information about what non-federal systems are used by employees. As a result, we continue to believe our recommendation, as written, is valid. In addition, the U.S. Postal Service concurred with our recommendation to assess the risks associated with non-federal systems.

Two of the eight entities that received recommendations emailed their comments. Specifically, the U.S. Capitol Police concurred with our recommendations and provided technical comments, which we incorporated as appropriate. In addition, the Department of Justice concurred with our recommendations and provided technical comments, which we incorporated as appropriate. Of note, the Department of Justice provided written and emailed comments.

We did not make recommendations to 13 of the 21 entities. Two of the 13 entities provided technical comments, which we incorporated as appropriate. Specifically, the Department of Commerce provided technical comments from the National Institute of Standards and Technology, and the National Aeronautics and Space Administration provided technical comments. The remaining 11 entities informed us that they had no comments.

We also provided a draft of this product for comment to two entities that did not receive our survey—the Office of Management and Budget and Clearview AI. The Office of Management and Budget informed us they had no comments, and Clearview AI provided technical comments, which we incorporated as appropriate.

As agreed with your offices, unless you publicly announce the contents of this report earlier, we plan no further distribution until 30 days from the report date. At that time, we will send copies to the relevant federal departments and entities.[34]

Gretta L. Goodwin
Director
Homeland Security and Justice

Appendix I: Objectives, Scope, and Methodology

This chapter examines (1) what federal agencies that employ law enforcement officers own and use facial recognition technology, (2) the type of activities these federal agencies use facial recognition technology to support, and (3) the extent that these federal agencies track employee use of facial recognition technology owned by non-federal entities, including state, local, tribal, territorial, and non-government entities.

This chapter is a public version of a sensitive report that we issued in April 2021.[35] Some federal agencies deemed information in our April report to be sensitive, which must be protected from public disclosure. Therefore, this chapter omits sensitive information about agency ownership and use of facial recognition technology. Although the information provided in this chapter is more limited, the report addresses the same objectives as the sensitive report and uses the same methodology.

To address all three objectives, we surveyed 42 federal agencies that employ law enforcement officers. Consistent with our prior work, we define federal law enforcement officers as full-time employees with federal arrest authority and who are authorized to carry firearms while on duty. To identify which agencies employ federal law enforcement officers, we reviewed the Bureau of Justice Statistics' 2016 Census of Federal Law Enforcement Officers.[36] The 2016 census identified the number of law enforcement officers

---

[34] Specifically, we will send copies to the 21 federal departments and other entities included in Table 6 at appendix I. In addition, we will send a copy to the Office of Management and Budget.

[35] GAO, *Facial Recognition Technology: Federal Law Enforcement Agencies Should Better Assess Privacy and Other Risks*, GAO-21-243SU (Washington, D.C.: April 28, 2021).

[36] Bureau of Justice Statistics, *Federal Law Enforcement Officers, 2016 – Statistical Tables*, NCJ 251922 (Washington, D.C.: October 2019).

employed by federal entities, with the exception of officers in the U.S. Armed Forces, officers stationed in foreign countries, and officers at the Central Intelligence Agency or Transportation Security Administration's Federal Air Marshal Service.

Our scope included 42 of the 86 federal entities identified in the 2016 census as employing law enforcement officers (see Table 6). The 2016 Census included Offices of Inspectors General; however, we excluded these agencies. We contacted the Council of the Inspectors General on Integrity and Efficiency, who stated that they were unaware of any Offices of Inspectors General that use facial recognition technology. However, we did include the Transportation Security Administration because it employs federal law enforcement officers, and was excluded from the 2016 census report for sensitivity reasons.

Table 6. 42 Federal Agencies Selected in GAO's Work

Federal Agency	Department
Forest Service	Agriculture
Bureau of Industry and Security	Commerce
National Oceanic and Atmospheric Administration, Office of Law Enforcement	Commerce
Office of Security	Commerce
Secretary's Protective Detail	Commerce
Pentagon Force Protection Agency	Defense
National Nuclear Security Administration	Energy
Food and Drug Administration, Office of Criminal Investigations	Health and Human Services
National Institutes of Health, Division of Police	Health and Human Services
U.S. Customs and Border Protection	Homeland Security
Federal Emergency Management Agency, Mount Weather Police	Homeland Security
Federal Protective Service	Homeland Security
U.S. Immigration and Customs Enforcement	Homeland Security
Office of the Chief Security Officer	Homeland Security
U.S. Secret Service	Homeland Security
Transportation Security Administration	Homeland Security
Bureau of Indian Affairs, Office of Justice Services	Interior
Bureau of Land Management	Interior
Bureau of Reclamation	Interior
U.S. Fish and Wildlife Service	Interior
U.S. Park Police	Interior
National Park Service Rangers	Interior
Bureau of Alcohol, Tobacco, Firearms and Explosives	Justice
Drug Enforcement Administration	Justice

Federal Bureau of Investigation	Justice
Federal Bureau of Prisons	Justice
U.S. Marshals Service	Justice
Division of Protective Operations	Labor
Bureau of Diplomatic Security	State
Bureau of Engraving and Printing Police	Treasury
Internal Revenue Service, Criminal Investigation Division	Treasury
U.S. Mint Police	Treasury
Police Service	Veterans Affairs
Amtrak Police Department	–
Federal Agency	Department
Environmental Protection Agency, Criminal Investigation Division	–
National Aeronautics and Space Administration, Office of Protective Services	–
U.S. Postal Inspection Service	–
Smithsonian Institution, Office of Protection Services	–
Tennessee Valley Authority Police	–
Administrative Office of the U.S. Courts, U.S. Probation and Pretrial Services	–
U.S. Capitol Police	–
Government Publishing Office, Uniform Police Branch	–

Legend: – Not Applicable.
Source: GAO information. | GAO-21-518.

To answer our first and second objectives, we administered a survey questionnaire to each of these 42 federal agencies. We administered the questionnaire by email from April through November 2020. Because we surveyed and obtained responses from all 42 agencies in the population defined by our scope, the summary results describing this group are not subject to errors from sampling and nonresponse. However, the practical difficulties of conducting any questionnaire survey may introduce other errors. For example, difficulties in how a particular question is interpreted by respondents, in the sources of information that are available to respondents, or in how we processed and analyzed the responses we received can influence the accuracy of the survey results. We took steps in the development of the questionnaire, the data collection, and the data analysis to minimize these potential errors, and to help ensure the accuracy of the answers that were obtained. We conducted pre-tests with three agencies in different departments to test the survey's applicability to a variety of facial recognition technology use cases, and revised the questionnaire based on the pre-tests.

When agencies submitted survey responses, we conducted an initial review for completeness, discrepancies, or logical errors within the responses, or discrepancies based on our prior knowledge (e.g., based on our review of a privacy impact assessment). We asked agencies to re-submit or clarify responses if necessary. We also confirmed our understanding of agency-provided information in semi-structured, open-ended follow-up interviews and information requests to the 20 agencies that reported they owned or used facial recognition technology. To help corroborate the information agencies provided in the questionnaire, we conducted a search of government contracting information, agency websites, and privacy documentation. When we discovered discrepancies, we followed up with the agency as appropriate.

We sent the questionnaire to audit liaisons or their designees for dissemination to relevant subject matter experts. We instructed these liaisons to provide a response on behalf of their organization, which would include all the agency's offices, employees, and contractors. The questionnaire defined facial recognition technology as a type of automated or semi-automated biometric technology that uses images for verification, identification, and/or investigative purposes. The questionnaire noted that facial recognition technology can be used for a variety of applications, such as verifying the identity claimed by an individual, identifying if an unknown individual exists in a gallery of known people, or comparing an unknown person to a gallery of known people to develop an investigative lead. In addition, we stated that a system with facial recognition technology may include a facial recognition algorithm, hardware, software, and a photo database. We asked agencies to include all uses of facial recognition technology in their response except for facial recognition technology that was solely used to authenticate the identity of the agency's employees and contractors to log into computers and phones.

The questionnaire asked whether at any point during January 2015 through March 2020, an agency owned a system with facial recognition technology, including systems in the process of being developed.[37] We used this time frame because March 2020 was the most recent full month for which information was available when we issued our questionnaire. Also, using a 5-year period allowed us to identify technology that was recently developed but not put in operation, and identify trends in facial recognition search data. We also asked whether at any point from April 2018 through March 2020, an agency's offices, employees, or contractors (1) accessed a system owned or operated by another entity, or (2) requested that another entity use its system to conduct a facial recognition search on their behalf.[38] All 42 agencies

[37] In our questionnaire, we stated that the term "own" includes systems that were procured or developed by the respective entity.

[38] For the purposes of this report, by saying an agency "used" another entity's system, we mean that an agency's offices, employees, and contractors (1) accessed a system owned or operated

responded to questions about the use of another entity's systems; however, some indicated that they could not guarantee the accuracy of the answers because they did not track this information. In some instances, employees and contractors had to work from their memory on the usage of another entity's systems. Agencies expressed this concern when we pretested the questionnaire, thus, we gathered this information from April 2018 (a 2-year period) instead of January 2015 (a 5-year period).

Agencies that owned a system completed an attachment with additional questions, such as the reason the agency used the system and the operational status of the system. In addition, we asked about the number of searches conducted on the system from 2015 through 2019, and number of photos included within the system as of March 31, 2020. We asked agency officials about the source of the data, whether the data were approximations or exact, whether definitions were consistently used in producing the data, and if there were any other data limitations that we should consider when reporting the information.

We requested additional information, through interviews and written requests, from agencies that reported in their questionnaire that they owned or used facial recognition technology. For example, if an agency reported having a system in operation, we requested privacy impact assessments and system descriptions. The information was used to help develop the system description and status throughout the report, including the detailed system appendices. In addition, we requested that 17 federal agencies provide information regarding whether they had used facial recognition technology on images of the civil unrest, riots, or protests from May through August 2020—the most recent full month for which information was available when we sent our questions. We also asked these 17 federal agencies whether they conducted facial recognition searches from January 6 through January 22, 2021 on images of the riot and civil unrest that occurred at the U.S. Capitol complex on January 6, 2021. We requested this information from 17 agencies that reported, via our questionnaire described above, they (1) had a system with facial recognition technology that was in operation, or (2) had used another entities' system. One exception was the Federal Bureau of Prisons, which we excluded because its system is used by employees to enter secure rooms.

To answer our third objective, we reviewed statutes and regulations, such as the Privacy Act of 1974. In addition, we interviewed or requested

---

by another entity, or (2) requested that another entity use its system to conduct a facial recognition search on their behalf.

information from officials from 14 agencies that reported using (1) non-federal systems, and (2) facial recognition technology to support criminal investigations. We determined that the control activities component of internal control was significant to this objective, along with the underlying principles that management should design control activities to achieve its objective and respond to risks. Specifically, we asked officials from the 14 agencies about their processes for gathering information on what non-federal systems are used by employees, and compared this information against our risk management framework and key aspects of *Standards for Internal Control in the Federal Government* (Principles 7, 10, and 16).[39]

We also interviewed and requested information from three additional federal agencies: Department of State's Bureau of Consular Affairs, Department of Homeland Security's Office of Biometric and Identity Management, and Department of Defense's Defense Forensic Science Center. We selected these agencies because multiple federal agencies we surveyed reported using these agencies' systems. We also met with other stakeholders to discuss law enforcement use of facial recognition technology, including other government entities, privacy advocacy groups, and non-government facial recognition technology providers. For example, we met with the Department of Commerce's National Institute of Standards and Technology, Georgetown Law Center on Privacy and Technology, and the International Biometrics and Identity Association.

While we did not comprehensively assess the reliability of data provided by agencies and other stakeholders, we took the steps described earlier and considered limitations identified by the agencies. We determined that data elements we assessed were sufficiently reliable for the purposes of this chapter.

The performance audit upon which this chapter is based was conducted from August 2019 to April 2021 in accordance with generally accepted government auditing standards. Those standards require that we plan and perform the audit to obtain sufficient, appropriate evidence to provide a reasonable basis for our findings and conclusions based on our audit objectives. We believe that the evidence obtained provides a reasonable basis for our findings and conclusions based on our audit objectives. We subsequently worked with the relevant entities from April 2021 to June 2021 to prepare this version of the original sensitive report for public release. This public version was also prepared in accordance with these standards.

---

[39] See GAO-14-704G and GAO-17-63.

Appendix II: Systems with Facial Recognition Technology Owned by Federal Agencies that Employ Law Enforcement Officers

Eight of the 42 surveyed agencies reported owning systems with facial recognition technology. Eight summaries are presented below—that is, one for each of the eight agencies.[40] Each summary includes a description of the agency's mission, an overview of the agency's ownership and use of other entities' systems, and detailed descriptions of each system owned by the agency. Information in these summaries was provided by the respective agency. See Figure 4 for an illustration of the layout of the summaries, including a description of each section in the summaries.

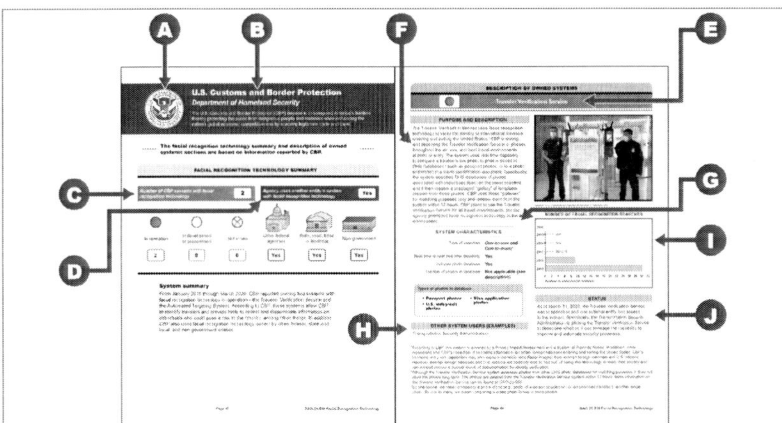

- **A** Agency seal.
- **B** Agency name, department name, and agency's mission
- **C** Number of systems owned or in procurement process at any point during January 1, 2015 through March 31, 2020. In addition, the status of these systems as of March 31, 2020.
- **D** Description of whether agency used a system owned by another entity at any point during April 1, 2018 through March 31, 2020.
- **E** System name and operational status as of March 31, 2020.
- **F** Purpose and Description: Description of why agency has (or is developing or procuring) the system and how agency uses (or plans to use) the system.

G – I are only included for systems in operation as of March 31, 2020:

- **G** System Characteristics: Description of system search capabilities, real-time capabilities, photo database, and the number and type of photos in the database.
  - **One-to-one or one-to-many.** By one-to-one, we mean comparing a probe photo (e.g., photo of a person of unknown or unconfirmed identity) to another single photo. By one-to-many, we mean comparing a probe photo to two or more photos.
  - **Real-time capability or near real-time capability.** By real-time capability, we mean the ability to conduct searches using live video streams. By near real-time capability, we mean the ability to conduct searches using very recently recorded video streams.
- **H** Other System Users (Examples): Examples of entities that had access to the system, or requested that the agency conduct a facial recognition search on their behalf, during January 1, 2015 through March 31, 2020.
- **I** Number of Facial Recognition Searches. Number of searches conducted on the system in calendar years 2015 through 2019. Some searches conducted on the system may not have been for law enforcement purposes.
- **J** Status: Description of system status as of March 31, 2020.

Source: GAO. | GAO-21-518.

Figure 4. Illustration of a Facial Recognition Technology Summary.

---

[40] This appendix omits information about systems with facial recognition technology deemed sensitive by the respective federal agency.

Federal Bureau of Investigation

Department of Justice

The Federal Bureau of Investigation (FBI) mission is to protect the American people and uphold the criminal laws and the Constitution of the United States. According to the FBI, its priorities include protecting the United States from terrorist attacks and against foreign intelligence operations and espionage, and cyber operations. In addition, agency priorities include combatting significant cyber-criminal activity, public corruption, and violent crime.

The facial recognition technology summary and description of owned systems sections are based on information reported by the FBI.

FACIAL RECOGNITION TECHNOLOGY SUMMARY

Number of FBI systems with facial recognition technology			*Agency uses another entity's system with facial recognition technology*		
6			Yes		
In operation	In development or procurement	Not in use	Other federal agencies	State, local, tribal, or territorial	Non-government
1	2	3	Yes	Yes	Yes

System summary

From January 2015 through March 2020, the FBI owned six systems with facial recognition technology; however, only one of these systems was in operation as of March 31, 2020. The FBI also used facial recognition technology owned by other federal, state and local, and non-government entities.

DESCRIPTION OF OWNED SYSTEMS

Next Generation Identification Interstate Photo System

PURPOSE AND DESCRIPTION

The Next Generation Identification Interstate Photo System allows authorized law enforcement officials to conduct facial recognition searches using a repository of approximately 42.8 million photos.[2] To conduct a search, a user uploads a probe photo and the facial recognition software compares the probe photo against the repository to find likely matches. The likely matches are returned in a ranked candidate list of two to 50 photos, depending on the user's specification. From 2015 through 2019, the total number of facial recognition searches conducted on this system was 232,915. FBI officials reported that the system has assisted investigations of credit card and identity fraud, missing persons, bank robberies, and violent crimes, among others.

Source: Federal Bureau of Investigation | GAO-21-518

SYSTEM CHARACTERISTICS

Type of searches:	**One-to-many**[3]
Real-time or near real-time capability:	**No**
Includes photo database:	**Yes**
Number of photos in database:	**42.8 million**

Types of photos in database

- **Mugshots**

NUMBER OF FACIAL RECOGNITION SEARCHES

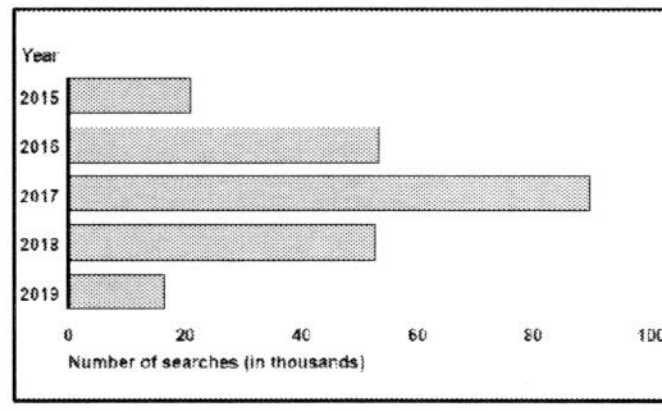

OTHER SYSTEM USERS (EXAMPLES)

U.S. Customs and Border Protection and state agencies within Arkansas, Arizona, Colorado, Delaware, Hawaii, Iowa, Kentucky, Louisiana, Maine, Michigan, New Mexico, Oklahoma, South Carolina, Texas, and West Virginia.

STATUS

As of March 31, 2020, the Next Generation Identification Interstate Photo System was in operation.

[2]According to the FBI, this system is covered by a Privacy Impact Assessment and a System of Records Notice.
[3]By one-to-many, we mean comparing a probe photo (e.g., photo of a person of unknown or unconfirmed identity) to two or more photos.

Federal Bureau of Investigation

Horus

PURPOSE AND DESCRIPTION

FBI is using Horus, a government-off-the-shelf product, for research purposes. Specifically, FBI forensic examiners compare two images to determine whether faces within the images represent the same individual (i.e., one-to-one comparisons). The FBI is researching Horus to determine whether this type of system could be incorporated into the FBI's one-to-one comparisons process. For example, researching whether the FBI could combine the use of forensic examiners and a facial recognition system to improve the accuracy of one-to-one comparisons. The system was also used in FBI educational settings to demonstrate how facial recognition technology works.

STATUS

As of March 31, 2020, the system was being tested but was not in operation.

Rank One

PURPOSE AND DESCRIPTION

FBI is using Rank One, a government-off-the-shelf product, for research purposes. Specifically, the FBI is researching Rank One to determine whether this type of system could be incorporated into the FBI's one-to-one comparisons process. For example, researching whether the FBI could combine the use of forensic examiners and a facial recognition system to improve the accuracy of one-to-one comparisons. The system was also used in FBI educational settings to demonstrate how facial recognition technology works.

STATUS

As of March 31, 2020, the system was being tested but was not in operation.

Automatic Face Detection and Recognition/Cluster Base

PURPOSE AND DESCRIPTION

FBI used Automatic Face Detection and Recognition/Cluster Base, a government-off-the-shelf product, for research purposes and for educational purposes in demonstrating facial recognition technology.

STATUS

As of March 31, 2020, the system was not in operation and no longer used for research purposes.

Federal Bureau of Investigation

Camera with Facial Recognition Software

PURPOSE AND DESCRIPTION

In 2015, the FBI purchased a camera that came with facial recognition software. The camera is used to take driver's license and passport photographs for undercover operations. The equipment is solely used for this purpose and the FBI has never used the facial recognition capability that was provided with the camera.

STATUS

As of March 31, 2020, the facial recognition technology software was not in operation. According to the FBI, it has no plans to use the software.

NeoFace Reveal

PURPOSE AND DESCRIPTION

FBI used NeoFace Reveal, a commercial-off-the-shelf product, for research purposes. Specifically, the FBI was researching NeoFace Reveal to determine whether this type of system could be incorporated into the FBI's one-to-one comparisons process. For example, researching whether the FBI could combine the use of forensic examiners and a facial recognition system to improve the accuracy of one-to-one comparisons. The system was also used in FBI educational settings to demonstrate how facial recognition technology works.

STATUS

As of March 31, 2020, the system was not in operation and no longer undergoing testing.

Federal Bureau of Prisons

Department of Justice

The Federal Bureau of Prisons (BOP) mission is to protect society by confining offenders in the controlled environments of prisons and community-based facilities that are safe, humane, cost-efficient, and appropriately secure, and that provide work and other self-improvement opportunities to assist offenders in becoming law-abiding citizens.

The facial recognition technology summary and description of owned systems sections are based on information reported by BOP.

FACIAL RECOGNITION TECHNOLOGY SUMMARY

Number of BOP systems with facial recognition technology **1**

Agency uses another entity's system with facial recognition technology **No**

In operation	In development or procurement	Not in use	Other federal agencies	State, local, tribal, or territorial	Non-government
1	0	0	No	No	No

System summary

From January 2015 through March 2020, the BOP owned one system with facial recognition technology that was in operation. BOP did not use facial recognition technology owned by other federal, state or local, and non-government entities.

U.S. Customs and Border Protection

Department of Homeland Security

The U.S. Customs and Border Protection (CBP) mission is to safeguard America's borders thereby protecting the public from dangerous people and materials while enhancing the nation's global economic competitiveness by enabling legitimate trade and travel.

The facial recognition technology summary and description of owned systems sections are based on information reported by CBP.

FACIAL RECOGNITION TECHNOLOGY SUMMARY

Number of CBP systems with facial recognition technology	2
In operation	2
In development or procurement	0
Not in use	0

Agency uses another entity's system with facial recognition technology	Yes
Other federal agencies	Yes
State, local, tribal, or territorial	Yes
Non-government	Yes

System summary

From January 2015 through March 2020, CBP reported owning two systems with facial recognition technology in operation—the Traveler Verification Service and the Automated Targeting System. According to CBP, these systems allow CBP to identify travelers and provide tools to collect and disseminate information on individuals who could pose a risk to the country, among other things. In addition, CBP also used facial recognition technology owned by other federal, state and local, and non-government entities.

DESCRIPTION OF OWNED SYSTEMS

Traveler Verification Service

PURPOSE AND DESCRIPTION

The Traveler Verification Service uses facial recognition technology to verify the identity of international travelers entering and exiting the United States.[6] CBP is testing and deploying the Traveler Verification Service in phases throughout the air, sea, and land travel environments at ports of entry. The system uses real-time capability to compare a traveler's live photo to photos stored in DHS databases,[7] such as passport photos, or to a photo embedded in a travel identification document. Specifically, the system searches DHS databases of photos associated with individuals listed on the travel manifest, and it then creates a prestaged "gallery" of templates created from those photos. CBP uses these "galleries" for matching purposes only and deletes them from the system within 12 hours. CBP plans to use the Traveler Verification Service for all travel environments, but the agency prioritized facial recognition technology in the air environment.

Source: U.S. Customs and Border Protection | GAO-21-518

SYSTEM CHARACTERISTICS

Type of searches:	**One-to-one and One-to-many**[8]
Real-time or near real-time capability:	**Yes**
Includes photo database:	**Yes**
Number of photos in database:	**Not applicable (see description)**

NUMBER OF FACIAL RECOGNITION SEARCHES

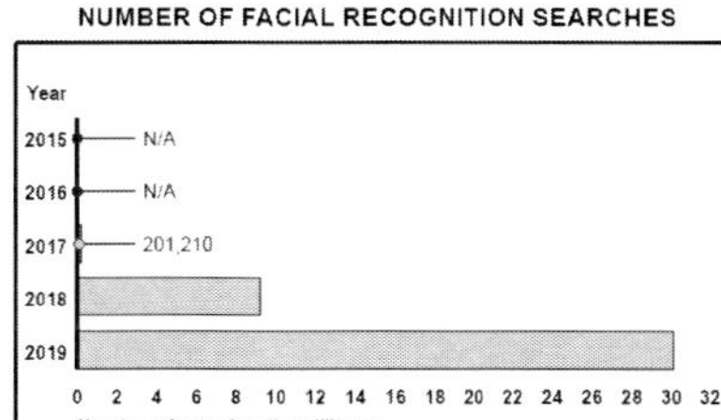

Types of photos in database

- **Passport photos**
- **U.S. entry/exit photos**
- **Visa application photos**

OTHER SYSTEM USERS (EXAMPLES)

Transportation Security Administration

STATUS

As of March 31, 2020, the Traveler Verification Service was in operation and one external entity had access to the system. Specifically, the Transportation Security Administration is piloting the Traveler Verification Service to determine whether it can leverage the capability to improve and automate security processes.

[6]According to CBP, this system is covered by a Privacy Impact Assessment and a System of Records Notice. In addition, while regulations limit CBP's collection of biometric information to certain foreign nationals entering and exiting the United States, CBP's biometric entry-exit capabilities may also capture biometric data (facial images) from exempt foreign nationals and U.S. citizens. However, exempt foreign nationals and U.S. citizens are routinely able to "opt out" of using this technology to verify their identity and can instead choose a manual check of documentation for identity verification.

[7]Although the Traveler Verification Service system accesses photos from other DHS photo databases for matching purposes, it does not store the photos long-term. The photos are deleted from the Traveler Verification Service system within 12 hours. More information on the Traveler Verification Service can be found at GAO-20-568.

[8]By one-to-one, we mean comparing a probe photo (e.g., photo of a person of unknown or unconfirmed identity) to another single photo. By one-to-many, we mean comparing a probe photo to two or more photos.

Automated Targeting System

PURPOSE AND DESCRIPTION

CBP uses the Automated Targeting System to facilitate trade and travel while managing the threat posed by individuals and cargo entering and exiting the United States.[9] Specifically, CBP uses the system for the following populations: (1) individuals seeking to enter or exit the U.S. whose names appear on travel manifests (e.g., flight or vessel manifests); (2) individuals applying for CBP programs facilitating travel to the U.S.; and (3) subjects of interest who require additional research and analysis. CBP will match photographs for these three populations against a predetermined gallery of photographs.

Source: U.S. Customs and Border Protection | GAO-21-518

SYSTEM CHARACTERISTICS

Type of searches:	**One-to-one and one-to-many**[10]
Real-time or near real-time capability:	**Yes**
Includes photo database:	**Yes**
Number of photos in database:	**15.5 million**

NUMBER OF FACIAL RECOGNITION SEARCHES

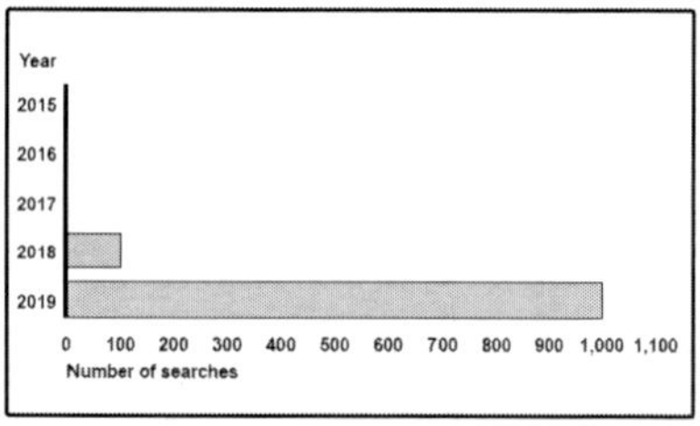

Types of photos in database

- **Passport photos**
- **U.S. entry/exit photos**
- **Visa application photos**
- **Mugshots**
- **State identification photos**
- **Video/Closed circuit television**

OTHER SYSTEM USERS (EXAMPLES)

None

STATUS

As of March 31, 2020, the Automated Targeting System was in operation.

[9]According to CBP, this system is covered by a Privacy Impact Assessment and a System of Records Notice.

[10]By one-to-one, we mean comparing a probe photo (e.g., photo of a person of unknown or unconfirmed identity) to another single photo. By one-to-many, we mean comparing a probe photo to two or more photos.

Transportation Security Administration

Department of Homeland Security

The Transportation Security Administration (TSA) mission is to protect the nation's transportation systems (e.g., aviation, rail, and public transit) to ensure freedom of movement for people and commerce.

The facial recognition technology summary and description of owned systems sections are based on information reported by TSA.

FACIAL RECOGNITION TECHNOLOGY SUMMARY

Number of TSA systems with facial recognition technology	2
Agency uses another entity's system with facial recognition technology	Yes

In operation	In development or procurement	Not in use	Other federal agencies	State, local, tribal, or territorial	Non-government
0	2	0	Yes	No	No

System summary

From January 2015 through March 2020, TSA did not have systems with facial recognition technology in operation. However, during this period, the agency began piloting two systems—the Credential Authentication Technology-2 (CAT-2) and the Automated Credential Authentication Technology (AutoCAT) system. TSA is also partnered with the U.S. Customs and Border Protection (CBP) on a number of multi-phased pilots using the Traveler Verification Service, which is a system owned by CBP. Specifically, TSA is testing biometric solutions to automate identity verification functions, which are currently performed manually at TSA checkpoints.

DESCRIPTION OF OWNED SYSTEMS

Transportation Security Administration

Credential Authentication Technology-2 (CAT-2)

PURPOSE AND DESCRIPTION

TSA is integrating one-to-one facial matching with existing Credential Authentication Technology (CAT) machines, and referring to these machines as CAT-2. TSA will use the CAT-2 solution to verify the identity of travelers who volunteer to use the technology. Specifically, the volunteer traveler steps up to the CAT-2 machine and inserts their identification document (e.g., driver's license). The CAT-2 then takes a photo of the traveler. Next, CAT-2 will match the photos with the image of the traveler's identification document using one-to-one facial matching, and a TSA officer will direct the traveler to the appropriate screening lane at the TSA checkpoint. In September 2019, TSA conducted a pilot of the CAT-2 units at McCarran International Airport TSA checkpoints. According to TSA officials, the pilot resulted in findings that will help inform future TSA plans and biometric requirements development, in addition to identifying and mitigating any performance issues and operations concerns. In response to the COVID-19 pandemic, TSA redesigned the solution to add a self-service component that allows passengers to scan their own identification document, which according to the agency, can promote social distancing.

STATUS

As of March 31, 2020, CAT-2 was in development. TSA began conducting a demonstration of CAT-2 at Reagan National Airport in August 2020. At the time, TSA was also planning to conduct field test events at Reagan National Airport, Miami International Airport, Phoenix Sky Harbor International Airport, Indianapolis International Airport, and a demonstration at Denver International Airport.

Automated Credential Authentication Technology (AutoCAT)

PURPOSE AND DESCRIPTION

TSA is studying AutoCAT security gates to determine whether they can, among other things, validate a traveler's identity using one-to-one facial matching. Specifically, when a volunteer traveler approaches a closed AutoCAT gate, it will scan the traveler's identification documents and take a photo of the traveler. Next, the AutoCAT will compare the live photo to the identification document or passport photo using one-to-one facial matching. If the photos match, the AutoCAT gate will open automatically and the traveler can pass through the security checkpoint. In February 2018, TSA conducted a 3-week demonstration of a biometric e-Gate at Los Angeles International Airport TSA checkpoints. According to TSA officials, the demonstration helped inform the design of future prototypes and requirements.

STATUS

As of March 31, 2020, AutoCAT was in development. TSA is validating system requirements and procuring prototypes and does not have additional demonstrations scheduled at the time.

U.S. Secret Service

Department of Homeland Security

The U.S. Secret Service (Secret Service) carries out the integrated missions of protection and investigations. According to the Secret Service, its central role is in the protection of the nation's leaders and the financial and critical infrastructure of the United States.

The facial recognition technology summary and description of owned systems sections are based on information reported by the Secret Service.

FACIAL RECOGNITION TECHNOLOGY SUMMARY

Number of Secret Service systems with facial recognition technology	1	*Agency uses another entity's system with facial recognition technology*	Yes

In operation	In development or procurement	Not in use	Other federal agencies	State, local, tribal or territorial	Non-government
0	0	1	Yes	Yes	Yes

System summary

From January 2015 through March 2020, Secret Service reported having one system with facial recognition technology. Secret Service deployed this system as part of the Facial Recognition Pilot within the White House Complex. Following the pilot, Secret Service did not implement the technology. The Secret Service also used facial recognition technology owned by other federal, state and local, and non-government entities.

DESCRIPTION OF OWNED SYSTEMS

U.S. Secret Service

Facial Recognition Pilot System

PURPOSE AND DESCRIPTION

According to the Secret Service, the Facial Recognition Pilot system was tested to determine whether it could be incorporated into the agency's White House Complex security operations. Specifically, the Secret Service stored photos of 23 volunteer employees within the Facial Recognition Pilot system. Next, as volunteers moved throughout the White House Complex, their images were captured by closed-circuit television cameras. In real time, the system compared the stored photos to images from the video footage to determine whether they represented the same individual.

STATUS

As of March 31, 2020, the Facial Recognition Pilot system was not in operation and no longer undergoing testing by the Secret Service. Agency officials told us that after six months of data collection and subsequent data analysis, they decided not to use the system in its operations and deleted all stored data and photos.

Police Service

Department of Veterans Affairs

The U.S. Department of Veterans Affairs (VA) Police Service's mission is to implement infrastructure protection, personal security, and law enforcement programs at VA facilities. VA Police Service also investigates serious incidents, conducts criminal analysis, and provides personal protection for the VA secretary and deputy secretary.

The facial recognition technology summary and description of owned systems sections are based on information reported by the VA Police Service.

FACIAL RECOGNITION TECHNOLOGY SUMMARY

Number of VA Police Service systems with facial recognition technology	4
Agency uses another entity's system with facial recognition technology	No

In operation	In development or procurement	Not in use	Other federal agencies	State, local, tribal, or territorial	Non-government
0	2	2	No	No	No

System summary

VA Police Service reported owning two systems with facial recognition technology from January 2015 through March 2020. However, neither system was put into operation. VA Police Service was also in the process of procuring two systems for use at medical centers in West Palm Beach, Florida and in North Chicago, Illinois, during the period of our review. VA Police Service intends to use the technology to, among other things, locate missing patients and identify individuals with a history of disruption.

DESCRIPTION OF OWNED SYSTEMS

VA Police Service

○ Motorola Avigilon Appearance Search

PURPOSE AND DESCRIPTION	STATUS
Motorola Avigilon Appearance Search is a commercially available system with facial and weapons recognition capabilities, according to the VA Police Service. The VA Police Service at the Lovell Federal Health Care Center in North Chicago, Illinois plans to use the technology to sort through hours of video to quickly locate a specific person, such as a missing patient, or a vehicle of interest. According to officials, this system will also allow VA Police Service to input a photo from a police bulletin (e.g., 'Be On the Lookout' bulletin), which would alert officers if someone suspected of an off-campus crime enters the medical center campus.	As of March 31, 2020, the VA Police Service at the Lovell Federal Health Care Center in North Chicago was in the process of procuring this system.

○ AnyVision

PURPOSE AND DESCRIPTION	STATUS
AnyVision is a commercially available system that can compare video footage to photos stored in a central database, according to the VA Police Service. The technology can search for suspects in video footage and display detections in a timeline. VA Police Service plans to use the software at the VA Medical Center at West Palm Beach to, among other things, respond to missing patients or those at risk of suicide, and to identify suspected or active threats.	As of March 31, 2020, the VA Medical Center at West Palm Beach was in the process of procuring this system to replace Veritone aiWARE.

⊗ Aventura

PURPOSE AND DESCRIPTION	STATUS
The VA Police Service at the Washington, DC Medical Center planned to use this facial recognition technology for surveillance purposes, among other things. For instance, the VA Police Service placed video cameras at the front and back entrance of the medical center, and could have used the facial recognition technology to alert officers of the arrival of a previously disruptive individual or to track missing patients.	As of March 31, 2020, this system was not in operation. VA Police Service did not perform any searches on this system, and deactivated it in 2017. According to VA Police Service, the agency still possesses the technology and could re-activate it if desired.

VA Police Service

⊗ Veritone aiWARE

PURPOSE AND DESCRIPTION	STATUS
The VA Medical Center at West Palm Beach has Veritone aiWARE, which is a web-based facial recognition system that can search for suspects in footage from all connected cameras, and display detections in a timeline, according to the VA Police Service. VA Police Service never implemented this system, but it had planned to use it to generate investigative leads and identify unknown individuals. For example, VA Police Service could have used Veritone aiWARE to alert officers of the arrival of a previously disruptive individual, or to track missing patients.	As of March 31, 2020, Veritone aiWARE is not in operation. VA Medical Center at West Palm Beach procured Veritone aiWARE, but according to VA officials, the technology became outdated due to delays in the agency's procurement process and delays related to the COVID-19 pandemic. In March 2020, VA Police Service began procuring AnyVision from the same vendor to replace this system.

Pentagon Force Protection Agency

Department of Defense

The Pentagon Force Protection Agency's (PFPA) mission is to protect and safeguard designated Department of Defense (DOD) personnel, resources, and facilities. Its duties include law enforcement, operational security, building surveillance, crisis prevention, counterintelligence and anti-terrorism, and protecting high-ranking DOD officials.

The facial recognition technology summary and description of owned systems sections are based on information reported by the PFPA.

FACIAL RECOGNITION TECHNOLOGY SUMMARY

Number of PFPA systems with facial recognition technology	*Agency uses another entity's system with facial recognition technology*
2	Yes

In operation	In development or procurement	Not in use	Other federal agencies	State, local, tribal, or territorial	Non-government
0	0	2	Yes	No	No

System summary

From January 2015 through March 2020, PFPA had two systems with facial recognition technology. According to officials, one of these systems was in operation as of March 31, 2020; however, the facial recognition technology component of the system was not in operation. The other system was tested once in 2019, but was not in operation as of March 31, 2020. PFPA has also requested facial recognition searches from the National Ground Intelligence Center; however, according to PFPA officials, they do not frequently request these searches.

DESCRIPTION OF OWNED SYSTEMS

Pentagon Force Protection Agency

Briefcam

PURPOSE AND DESCRIPTION

Briefcam is a live video synopsis tool that includes facial recognition technology, among other capabilities, according to PFPA. Specifically, PFPA operates approximately 3,000 cameras in the Washington, DC region, resulting in thousands of hours of video footage. According to PFPA officials, Briefcam can consolidate hours of video footage into minutes, which helps them review the footage. In addition, Briefcam has the capability to, among other things, conduct real-time facial recognition searches.

STATUS

As of March 31, 2020, the facial recognition technology component of the system was not in operation. Specifically, PFPA uses the video consolidation capabilities, but has not used the facial recognition technology accessible within Briefcam. According to PFPA officials, they do not plan to use Briefcam's facial recognition technology in the future.

Sirchie

PURPOSE AND DESCRIPTION

Sirchie software is part of a remote video system used by PFPA. The system is aboard a mobile command center that PFPA can deploy to areas lacking camera infrastructure (e.g., areas at a special event that lack cameras). A primary function of the system is sending information, such as captured video, back to a central command center. Moreover, the system is capable of housing a facial image database for facial recognition searches, according to PFPA.

STATUS

On one occasion, PFPA tested the facial recognition capability of Sirchie on four of its employees. However, as of March 31, 2020, the facial recognition technology within Sirchie was not in operation and a photo database was never uploaded. According to PFPA officials, they do not plan to use this system's facial recognition technology capability in the future.

Office of Protective Services

National Aeronautics and Space Administration

The Office of Protective Services provides security, fire, and medical services for National Aeronautics and Space Administration (NASA) facilities, personnel, and visitors.

The facial recognition technology summary and description of owned systems sections are based on information reported by the NASA Office of Protective Services.

FACIAL RECOGNITION TECHNOLOGY SUMMARY

Number of NASA Office of Protective Services systems with facial recognition technology	1	*Agency uses another entity's system with facial recognition technology*	No

In operation	In development or procurement	Not in use	Other federal agencies	State, local, tribal, or territorial	Non-government
0	0	1	No	No	No

System summary

From January 2015 through March 2020, NASA's Office of Protective Services participated in a proof of concept test that included the facial recognition technology FaceFirst. However, the technology was not deployed after it was tested.

DESCRIPTION OF OWNED SYSTEMS

NASA Office of Protective Services

FaceFirst

PURPOSE AND DESCRIPTION

NASA's Office of Protective Services worked with NASA's Technology and Innovation Labs to provide a proof of concept that included the facial recognition technology FaceFirst, a commercially available off-the-shelf product. The proof of concept sought to enhance the user experience at NASA's forgotten badge kiosk by applying a facial recognition solution, which can be faster and more reliable than matching a driver's license to database records, according to Office of Protective Services officials. In addition, the solution sought to eliminate the reliance on access tokens such as badges and enable more effective physical security.

STATUS

As of March 31, 2020, the technology was not in operation and was no longer being tested. According to agency officials, the cost to deploy the technology exceeded the budget, and there are no plans to deploy the technology.

Appendix III: Other Federal Systems with Facial Recognition Technology

This appendix includes facial recognition technology summaries for three federal agencies: the Department of State's Bureau of Consular Affairs,

Department of Defense's Defense Forensic Science Center, and the Department of Homeland Security's Office of Biometric Identity Management. These agencies were not included in our survey. However, we are presenting summary descriptions because they have facial recognition technology that multiple surveyed agencies reported using. Information in the summaries were provided by the respective agency. The summaries use a similar format to those included in Appendix II. See Figure 4 in Appendix II for an illustration of the layout of the summaries, including a description of each section in the summaries.

Bureau of Consular Affairs - Department of State

Integrated Biometric System

This system description is based on information reported by the Bureau of Consular Affairs.

PURPOSE AND DESCRIPTION

The Integrated Biometric System provides the Department of State's passport agencies and consular posts around the world with information to use in evaluating visa and passport applications (e.g. photos, region of residence or nationality, birth dates, and assigned identification number).[1] The system includes, among other things, facial recognition technology. State officials can use the facial recognition technology to, for example, verify an applicant's identity or to determine whether an individual has previously applied for a visa. The system is generally used for non-law enforcement purposes; however, some federal agencies employing law enforcement officers conduct facial recognition searches on the system. For example, the Bureau of Diplomatic Security at the Department of State can compare a photo of an unknown individual to photos stored in the Integrated Biometric System to generate investigative leads.

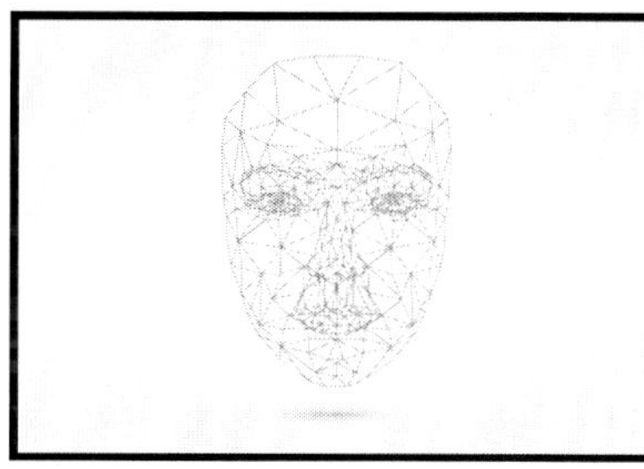

Source: lidiia/stock.adobe.com. | GAO-21-518

SYSTEM CHARACTERISTICS

Type of searches:	**One-to-many**[2]
Real-time or near real-time capability:	**No**
Includes photo database:	**Yes**
Number of photos in database:	**465 million**

Types of photos in database

- **Passport and visa application photos**
- **Terrorist Screening Database photos**

NUMBER OF FACIAL RECOGNITION SEARCHES

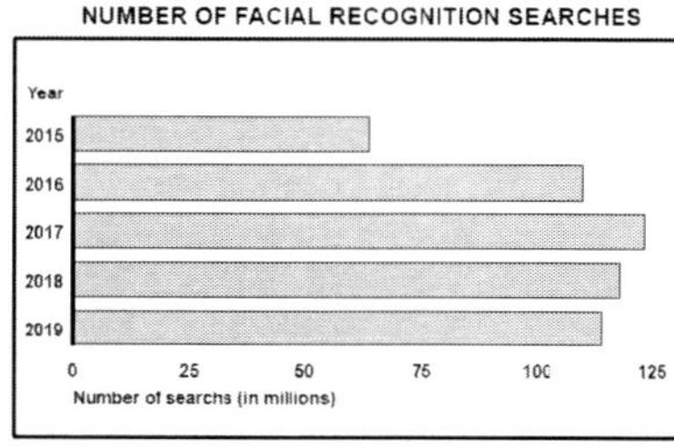

OTHER SYSTEM USERS (EXAMPLES)

U.S. Customs and Border Protection, U.S. Immigration and Customs Enforcement, and Federal Bureau of Investigation.[3]

STATUS

As of March 31, 2020, the Integrated Biometric System was in operation.

[1]According to the Bureau of Consular Affairs, this system is covered by a Privacy Impact Assessment and a System of Records Notice.
[2]By one-to-many, we mean comparing a probe photo (e.g., photo of a person of unknown or unconfirmed identity) to two or more photos.
[3]The Integrated Biometric System includes a passport photo gallery and a visa photo gallery, among others. While other agencies with access agreements can conduct their own facial recognition searches of visa galleries, only Department of State officials can access the passport gallery for facial recognition searches. In addition, the Department of State has an agreement with the Federal Bureau of Investigation (FBI) which allows the FBI to receive circumscribed results to searches done on their behalf by Department of State officials within the passport gallery.

Defense Forensic Science Center - Department of Defense

Automated Biometric Identification System (ABIS)

This system description is based on information reported by the Defense Forensic Science Center.

PURPOSE AND DESCRIPTION

The Department of Defense uses ABIS to identify known or suspected threat actors around the world.[4] The system has numerous capabilities, including facial recognition technology that may be used with the approval of a trained examiner. For example, the system's facial recognition technology can match photos and provide a confidence score about whether faces within the images represent the same individual. Department of Defense may use ABIS to, for example, vet individuals requesting access to military bases abroad.

Source: Defense Forensic Science Center | GAO-21-518

SYSTEM CHARACTERISTICS

Type of searches: **One-to-many**[5]

Real-time or near real-time capability: **No**

Includes photo database: **Yes**

Number of photos in database: **22.8 million**

Types of photos in database

- **Foreign nationals and U.S. citizens who are known or suspected threats**
- **Individuals requesting access to military bases abroad**

NUMBER OF FACIAL RECOGNITION SEARCHES

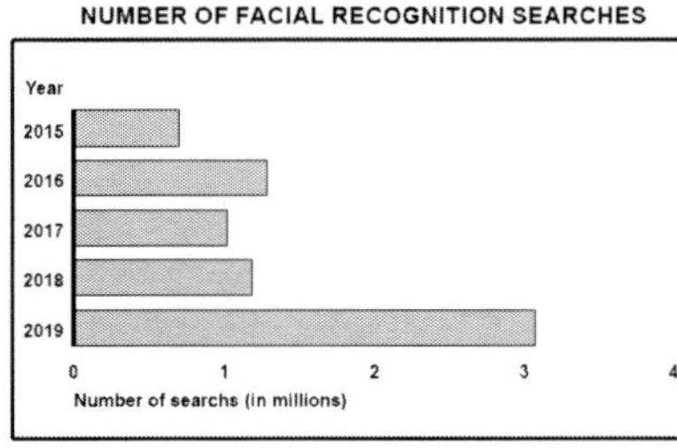

OTHER SYSTEM USERS (EXAMPLES)

Federal Bureau of Investigation, U.S. Customs and Border Protection, U.S. Immigration and Customs Enforcement, and Bureau of Diplomatic Security

STATUS

As of March 31, 2020, ABIS was in operation. According to officials, Department of Defense began using ABIS in 2004, and added facial recognition technology in 2009.

---

[4]According to the Department of Defense, this system is covered by a Privacy Impact Assessment and a System of Records Notice.

[5]By one-to-many, we mean comparing a probe photo (e.g., photo of a person of unknown or unconfirmed identity) to two or more photos.

Office of Biometric Identity Management (OBIM) - Department of Homeland Security

Automated Biometric Identification System (IDENT)

These system descriptions are based on information reported by the Office of Biometric Identity Management.

PURPOSE AND DESCRIPTION

IDENT, developed in 1994, is DHS's primary system for storing and processing biometric data (e.g., facial images and fingerprints) and corresponding biographic data (e.g., date of birth).[6] OBIM uses IDENT to offer three facial recognition services to partners, including federal law enforcement. These services allow partners to verify an individual's identity, determine whether an individual in two separate photos is the same person, and compare a probe photo against images stored in IDENT to create a list of potential candidates. For example, regarding the candidate search service, a partner can send OBIM a photo of an unknown individual. Next, the photo is uploaded into IDENT and compared against images within IDENT's photo repository. IDENT provides a list of potential matches (i.e., candidates), rather than a single match, for investigative leads only.

Source: Office of Biometric Identity Management | GAO-21-518

SYSTEM CHARACTERISTICS

Type of searches:	One-to-one and one-to-many[7]
Real-time or near real-time capability:	No
Includes photo database:	Yes
Number of photos in database:	836 million

Types of photos in database

- Mugshots
- Passport photos
- U.S. entry/exit photos
- Credential photos
- Visa application photos

NUMBER OF FACIAL RECOGNITION SEARCHES

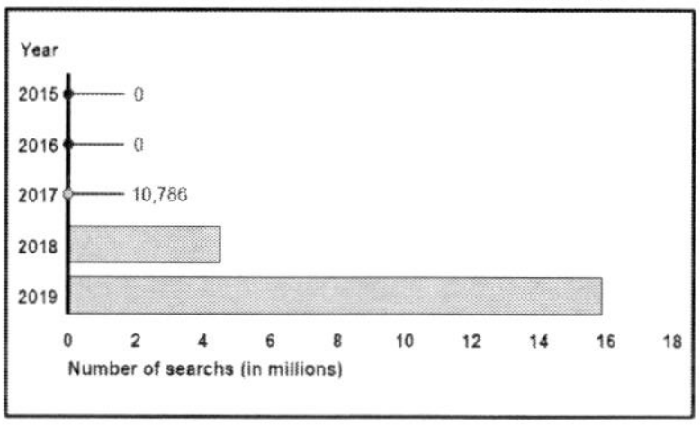

OTHER SYSTEM USERS (EXAMPLES)

U.S. Customs and Border Protection, U.S. Immigration and Customs Enforcement, Department of Homeland Security's Office of Intelligence and Analysis, and the National Counterterrorism Center

STATUS

As of March 31, 2020, IDENT was in operation. As discussed further below, DHS is developing a new system (i.e., HART), which will replace IDENT.

---

[6]According to OBIM, this system is covered by a Privacy Impact Assessment and a System of Records Notice.

[7]By one-to-one, we mean comparing a probe photo (e.g., photo of a person of unknown or unconfirmed identity) to another single photo. By one-to-many, we mean comparing a probe photo to two or more photos.

Office of Biometric Identity Management

Homeland Advanced Recognition Technology System (HART)

PURPOSE AND DESCRIPTION

OBIM is developing HART, which will replace IDENT as DHS's primary system for matching, storing, and sharing biometric information. According to OBIM, HART Increment 1 will initially replace IDENT's current capabilities with a more robust, flexible, and scalable system that better addresses current and future biometric needs. In addition, Increment 2 will deliver multimodal fusion, as well as additional biometric services and enhancements, according to OBIM. Finally, according to OBIM, future capability development will be driven by customer priorities, and will include additional biometric services and additional biometric modalities such as DNA and Palm matching among other things.

STATUS

As of March 2020, HART was in development. According to OBIM, as of March 2021, Increment 1 is expected to be delivered between the first and third quarters of 2022. Once OBIM deems Increment 1 a success, HART will become operational and IDENT will be decommissioned. Increment 2 is expected to be deployed shortly thereafter in 2022. Additional capabilities will be added in subsequent years.

Appendix IV: Comments from the Department of Health and Human Services

DEPARTMENT OF HEALTH & HUMAN SERVICES OFFICE OF THE SECRETARY

Assistant Secretary for Legislation
Washington, DC 20201

May 12, 2021

Gretta Goodwin
Director, Homeland Security and Justice
U.S. Government Accountability Office
441 G Street NW
Washington, DC 20548

Dear Ms. Goodwin:

Attached are comments on the U.S. Government Accountability Office's (GAO) report entitled, "*Facial Recognition Technology: Federal Law Enforcement Agencies Should Better Assess Privacy and Other Risks*" (Job code 105161/GAO-21-518).

The Department appreciates the opportunity to review this report prior to publication.

Sincerely,

Rose M. Sullivan -S Digitally signed by Rose M. Sullivan -S Date: 2021.05.25 16:08:24 -04'00'

Rose Sullivan
Acting Assistant Secretary for Legislation
Principal Deputy Assistant Secretary for Legislation

GENERAL COMMENTS FROM THE DEPARTMENT OF HEALTH & HUMAN SERVICES ON THE GOVERNMENT ACCOUNTABILITY OFFICE'S DRAFT REPORT ENTITLED — FACIAL RECOGNITION TECHNOLOGY: FEDERAL LAW ENFORCEMENT AGENCIES SHOULD BETTER ASSESS PRIVACY AND OTHER RISKS (GAO-21-518)

The U.S. Department of Health & Human Services (HHS) appreciates the opportunity from the Government Accountability Office (GAO) to review and comment on this draft report.

Recommendation 19
The Assistant Commissioner of the Food and Drug Administration's Office of Criminal Investigations should implement a mechanism to track what non-federal systems with facial recognition technology are used by employees to support investigative activities.

HHS Response
FDA concurs with this GAO recommendation, and the Food and Drug Administration's Office of Criminal Investigations (OCI) intends to implement a mechanism to track what federal and non-federal (including state, local, tribal, territorial, and non-governmental) facial recognition technology (FRT) systems are used by OCI agents to support investigative activities. OCI intends that such tracking mechanism will include the name of the FRT system, owner of the FRT system, and how use of the FRT system supported OCI's mission.

Recommendation 20
The Assistant Commissioner of the Food and Drug Administration's Office of Criminal Investigations should, after implementing a mechanism to track non-federal systems, assess the risks of using such systems, including privacy and accuracy related risks.

HHS Response
FDA concurs with GAO's recommendation to assess the risks of using non-federal FRT systems that are identified through OCI's tracking mechanism. OCI will assemble a list of potential risks posed by the use of non-federal FRT systems, including risks related to privacy issues as well as system accuracy-related risks. OCI will use this list to help FDA identify threats to achieving investigative goals and objectives, and if appropriate, assist FDA in the adoption of treatment strategies to manage priority risks.

Appendix V: Comments from the Department of Homeland Security

U.S. Department of Homeland Security
Washington, DC 20528

Homeland Security

May 11, 2021

Gretta L. Goodwin
Director, Homeland Security and Justice
U.S. Government Accountability Office
441 G Street, NW
Washington, DC 20548

Re: Management Response to Draft Report GAO-21-518, "FACIAL RECOGNITION TECHNOLOGY: Federal Law Enforcement Agencies Should Better Assess Privacy and Other Risks"

Dear Ms. Goodwin:

Thank you for the opportunity to comment on this draft report. The U.S. Department of Homeland Security (DHS or the Department) appreciates the U.S. Government Accountability Office's (GAO) work in planning and conducting its review and issuing this report.

The Department is pleased to note GAO's recognition that the U.S. Immigration and Customs Enforcement assessed privacy risks associated with its use of facial recognition technology, including non-federal systems, and that the agency identified privacy risks and possible actions to mitigate those risks. As the Department expands the use of facial recognition technology in its missions, DHS is committed to improved tracking of the use of non-Federal systems for mission purposes, as well as assessing the privacy and accuracy risks of utilizing such systems.

The draft report contained 24 recommendations, including four for DHS with which the Department concurs. Attached find our detailed response to each recommendation. DHS previously submitted technical comments addressing several accuracy, contextual, and other issues under a separate cover for GAO's consideration.

Again, thank you for the opportunity to review and comment on this draft report. Please feel free to contact me if you have any questions. We look forward to working with you again in the future.

Sincerely,

JIM H
CRUMPACKER

Digitally signed by JIM H CRUMPACKER
Date: 2021.05.11 16:37:37 -04'00'

JIM H. CRUMPACKER, CIA, CFE
Director
Departmental GAO-OIG Liaison Office

Attachment

Attachment: Management Response to Recommendations Contained in GAO-21-518

GAO recommended that the Commissioner of U.S. Customs and Border Protection (CBP):

Recommendation 9: Implement a mechanism to track what non-federal systems with facial recognition technology are used by employees to support investigative activities.

Response: Concur. Pursuant to CBP's Privacy Directive 2120-010, "Privacy Policy, Compliance, and Implementation," dated January 2, 2015, all CBP personnel are responsible for notifying the CBP Privacy Office regarding the implementation, or proposed implementation, of technologies that involve personally identifiable information (PII) or that may otherwise impact the privacy of individuals. While directive 2120-010 broadly defines PII, it does not explicitly indicate that facial recognition technologies are potentially privacy invasive tools. The Directive also requires that all programs using tools, systems, and technologies—or implementing a program, pilot, or rulemaking—must, in coordination with the CBP Privacy Office, complete a Privacy Threshold Analysis to determine whether PII is involved.

By the end of 2021, the CBP Privacy Office will update CBP's Privacy Directive to more explicitly identify the applicability of these reporting and coordination requirements with regard to the use of facial recognition technologies, regardless of whether they are owned or operated by CBP. Upon approval and implementation by the Commissioner of CBP, the Directive will be distributed to the CBP workforce. Additionally, the CBP Privacy Office will create and circulate messaging specific to the use and implementation of facial recognition technologies to the entirety of CBP's workforce. This increase in awareness should foster more active participation in CBP's already robust Privacy Compliance process, reinforcing its function as a mechanism for the tracking of CBP systems, programs, and pilot efforts. Estimated Completion Date (ECD): December 31, 2021.

Recommendation 10: After implementing a mechanism to track non-federal systems, assess the risks of using such systems, including privacy and accuracy related risks.

Response: Concur. CBP's Privacy Compliance process is substantial, and includes reviews of IT systems, technologies, rulemakings, programs, pilot projects, and other activities to determine what, if any, PII is collected, used, or shared. This assessment allows CBP to determine whether additional policy, technological, or physical safeguards are necessary to ensure that data is protected from unauthorized use or disclosure. Increased awareness and understanding of the requirement to coordinate with the CBP Privacy Office, brought on by updating and publishing the Privacy Directive, will present new opportunities for the review of use cases and tools associated with facial recognition technologies that the Privacy Office may not have been previously aware. However, the CBP Privacy Office will also conduct training and outreach for individuals who have the ability to procure access to these types of vendors and technologies, via a government purchase card, outside of the typical IT acquisition review process. ECD: December 31, 2021.

GAO recommended that the Director of the United States Secret Service (USSS):

Recommendation 11: Implement a mechanism to track what non-federal systems with facial recognition technology are used by employees to support investigative activities.

Response: Concur. The USSS Office of Investigations recognizes the importance of establishing visibility regarding the investigative tools utilized by agency personnel during the course of official investigative activities. Consequently, the Investigative Support Division and the Criminal Investigative Division are jointly working to enhance the functionality of the internal agency case management system to collect this information. Once established, the functional augmentation of the internal case management system will serve as a tracking mechanism for usage of non-federal facial recognition technologies. As such, the system will require USSS personnel to record the usage, or request for the usage, of non-federal facial recognition technology systems as an official investigative activity within the internal case management system. Specifically, the internal case management system will provide a mechanism to collect information about the use of facial recognition technologies used in support of investigative activities to include the: (1) owner and/or operator of the facial recognition technology system; (2) facial recognition technology system utilized; and (3) date the search was conducted. ECD: December 31, 2021.

Recommendation 12: After implementing a mechanism to track non-federal systems, assess the risks of using such systems, including privacy and accuracy related risks.

Response: Concur. Upon completion of the functional augmentation within the internal case management system, as well as implementation of the mechanism to capture usage of non-federal facial recognition technology systems, the USSS Office of Investigations will be better postured to assess the risks of using such systems. This improved visibility into the non-federal facial recognition technology used by agency personnel will enable the Office of Investigations to: (1) actively query usage of such systems; (2) ensure best practices for usage; and (3) assess the risks associated with privacy and/or accuracy deficiencies. Furthermore, the Office of Investigations will use the information garnered from this continual assessment to provide guidance to agency personnel as to appropriate usage of, or restrictions to the usage of, non-federal facial recognition technology systems. ECD: December 31, 2021.

Appendix VI: Comments from the Department of the Interior

United States Department of the Interior

OFFICE OF THE SECRETARY
Washington, DC 20240

Gretta L. Goodwin
Director, Justice and Law Enforcement Issues
Homeland Security and Justice Team
U.S. Government Accountability Office
441 G Street, NW
Washington, DC 20548

Dear Director Goodwin,

Thank you for providing the Department of the Interior (Department) an opportunity to review and comment on the draft Government Accountability Office (GAO) report entitled, *Facial Recognition Technology: Federal Law Enforcement Agencies Should Better Assess Privacy and Other Risks* (GAO-21-243SU). We appreciate GAO's review and feedback related to the federal law enforcement use of facial recognition technology.

The GAO issued four recommendations to the Department as part of its overall findings to improve these processes. The report contains two recommendations each for the U.S. Fish and Wildlife Service (FWS) and National Park Service (NPS). Below is a summary of actions taken or planned to implement the recommendations.

Recommendation 13: "The Director of the U.S. Fish and Wildlife Service should implement a mechanism to track what non-federal systems with facial recognition technology are used by employees to support investigative activities."

Response: Concur. The Assistant Director, Office of Law Enforcement, entered into a licensing agreement with Clearview Facial Recognition Technology (FRT) on July 1, 2020, to provide all facial recognition requests for the FWS law enforcement needs. The FWS has limited use to two licensed users. Clearview's reporting capabilities include tracking of all inquiries by license and case number. Clearview is the only FRT authorized for use in the FWS. Use of FRT during an investigation is entered into the case file in the Law Enforcement Management Information System (LEMIS). The Assistant Director will issue a Chief's Directive outlining the process for law enforcement personnel to request the use of facial recognition by the licensed users in cases requiring facial recognition.

Responsible Official: Assistant Director, Office of Law Enforcement

Target Date: March 1, 2022

1

Recommendation 14: "The Director of the U.S. Fish and Wildlife Service should, after implementing a mechanism to track non-federal systems, assess the risks of using such systems, including privacy and accuracy related risks."

Response: Concur. The Assistant Director, Office of Law Enforcement, will issue a Chief's Directive, in coordination with the Department's Office of the Chief Information Officer, as appropriate, requiring a risk assessment of any facial recognition technology the FWS may use, to include an assessment of privacy and accuracy risks.

Responsible Official: Assistant Director, Office of Law Enforcement

Target Date: March 1, 2022

Recommendation 15: "The Chief of the U.S. Park Police should implement a mechanism to track what non-federal systems with facial recognition technology are used by employees to support investigative activities."

Response: Concur. The Chief, U.S. Park Police, will establish a policy outlining a tracking mechanism for the use of non-federal systems with facial recognition technology used by its employees to support investigative activities.

Responsible Official: Chief, U.S. Park Police

Target Date: December 31, 2021

Recommendation 16: "The Chief of the U.S. Park Police should, after implementing a mechanism to track non-federal systems, assess the risks of using such systems, including privacy and accuracy related risks."

Response: Concur. The Chief, U.S. Park Police (USPP), will issue a policy, in coordination with the Department's Office of the Chief Information Officer, as appropriate, requiring a risk assessment of any facial recognition technology the USPP may use, to include an assessment of privacy and accuracy risks.

Responsible Official: Chief, U.S. Park Police

Target Date: March 1, 2022

If you have any questions or need additional information, please contact the Internal Control and Audit Follow-up Division of the Office of Financial Management at DOI_PFM_ICAF@ios.doi.gov.

Sincerely,

SHANNON ESTENOZ (Affiliate) Digitally signed by SHANNON ESTENOZ (Affiliate)

Shannon Estenoz
Principal Deputy Assistant Secretary for
Fish and Wildlife and Parks
Exercising the Delegated Authority of the Assistant
Secretary for Fish and Wildlife and Parks

Appendix VII: Comments from the Department of State

United States Department of State
Comptroller
Washington, DC 20520

MAR 10 2021

Thomas Melito
Managing Director
International Affairs and Trade
Government Accountability Office
441 G Street, N.W.
Washington, D.C. 20548-0001

Dear Mr. Melito:

We appreciate the opportunity to review your draft report, "FACIAL RECOGNITION TECHNOLOGY: Federal Law Enforcement Agencies Should Better Assess Privacy and Other Risks" GAO Job Code 103705.

The enclosed Department of State comments are provided for incorporation with this letter as an appendix to the final report.

Sincerely,

Jeffrey C. Mounts

Enclosure:
As stated

cc: GAO – Gretta Goodwin
DS – Todd J. Brown
OIG - Norman Brown

Department of State Comments on GAO Draft Report

FACIAL RECOGNITION TECHNOLOGY: Federal Law Enforcement Agencies Should Better Assess Privacy and Other Risks
(GAO-21-243SU, GAO Code 103705)

Thank you for the opportunity to comment on your draft report entitled *"Facial Recognition Technology: Federal Law Enforcement Agencies Should Better Assess Privacy and Other Risks."* The report includes two recommendations for the Department of State. The Department concurs with these recommendations.

Recommendation 17: The Assistant Secretary for Diplomatic Security (DS) should implement a mechanism to track what non-federal systems with facial recognition technology are used by employees to support investigative activities.

Response: The Department of State agrees. The Department, through the Bureau of Diplomatic Security (DS), supports the need for tracking and managing access to "non-federal systems with facial recognition technology" and the use of these systems in support of DS criminal investigations.

Prior to this study and concurrent with it, DS has been developing internal controls and standard operating procedures to ensure that access to any "non-federal systems with facial recognition technology" by DS special agents and analysts is properly vetted and that accounts are managed centrally through Bureau system account management processes. These procedures and functions are in the final drafting phase and are intended to be implemented this fiscal year globally.

Recommendation 18: The Assistant Secretary for DS should, after implementing a mechanism to track non-federal systems, assess the risks of using such systems, including privacy and accuracy related risks.

Response: The Department of State agrees. As part of the implementation of the controls referred to in response to Recommendation 17, DS intends to establish an internal review panel to evaluate and review any "non-federal systems with facial recognition technology" that might be used. The panel would be intended to centralize contracting reviews and evaluate the provider's privacy assessments and practices, as well as the internal processes for data collection, in order to assess the risks of using such a system. DS also intends to centralize control over the contracts and funding to ensure users are not inadvertently encouraged by providers to utilize "non-federal systems with facial recognition technology" that lack approval and oversight.

The Department thanks the GAO for this constructive audit and will promptly implement the above recommendations.

Appendix VIII: Comments from the Department of the Treasury

DEPARTMENT OF THE TREASURY
INTERNAL REVENUE SERVICE
WASHINGTON, DC 20224

DEPUTY COMMISSIONER

March 10, 2021

James R. McTigue, Jr
Director, Tax Issues/Strategic Issues
United States Government Accountability Office
441 G Street, NW
Washington, DC 20548

Dear Mr. McTigue:

Thank you for the opportunity to review the draft report of the Government Accountability Office entitled "*Facial Recognition Technology: Federal Law Enforcement Agencies Should Better Assess Privacy and Other Risks*" (GAO-21-243SU). The focus of the report is to review how federal law enforcement agencies utilize facial recognition technology to assist in criminal investigations. We agree with the report and its findings.

This report includes two recommendations to each of twelve Agencies. These recommendations are designed to assist the Agencies in tracking what non-federal systems are used by employees, and to assess the risks of using these systems. With that as background, enclosed are comments on the draft report's recommendations directed to the IRS.

We appreciate having the opportunity to review and comment on the draft report. Responses to your specific recommendations are enclosed. If you have questions, please contact me, or a member of your staff may contact Guy Ficco, Executive Director of Operations, at 202-317-3804.

Sincerely,

Sunita B. Lough Digitally signed by Sunita B. Lough Date: 2021.03.10 17:13:23 -05'00'

Sunita Lough
Deputy Commissioner for Services and Enforcement

Enclosure

Enclosure

Comments on the GAO Recommendations Directed to the IRS

Recommendation 1:
The Chief of the Internal Revenue Service's Criminal Investigative Division should implement a mechanism to track what non-federal systems with facial recognition technology are used by employees to support investigative activities.

Comment:
The IRS agrees that the Chief of the Internal Revenue Service's Criminal Investigative Division (CI) should implement a mechanism to track non-federal facial recognition technology systems used by employees to support investigative activities. To that end, requests that would involve the use of a facial recognition technology system, if carried out by IRS CI's National Forensic Laboratory (NFL), would be tracked using an electronic Laboratory Information Management System (LIMS). LIMS is used for creating, collecting, and storing all case-related information, to include database searches and the results of those searches, as well as equipment and technology used in analysis and to obtain reportable results. Furthermore, CI's Technology Operations and Investigative Services (TOIS) uses cyber security tools to monitor software and systems accessed on the CI network. Therefore, the IRS has mechanisms in place capable of tracking non-federal facial recognition technology systems used by the IRS in support of investigative activities.

It should be noted that CI does not currently have the ability to directly access or connect to federal or non-federal systems with facial recognition technology capabilities.

Recommendation 2:
The Chief of the Internal Revenue Service's Criminal Investigative Division should, after implementing a mechanism to track non-federal systems, assess the risks of using such systems, including privacy and accuracy related risks.

Comment:
The IRS agrees that the Chief of the Internal Revenue Service's Criminal Investigative Division should assess the risks of using non-federal facial recognition technology systems, including privacy and accuracy related risks. To that end, IRS subject matter experts serve as members of professional organizations and groups such as the Facial Recognition Interagency Working Group. Active participation allows the IRS to continually assess risks of and best practices for using non-federal facial recognition technology systems. Also, documents such as the Bureau of Justice Assistance Face Recognition Policy Development Template are available to provide a framework under which a face recognition program can be operated and in a manner that complies with applicable laws, minimizes risks, and establishes accountability and oversight.

Although IRS does not currently possess or is not connected to facial recognition technology systems, IRS subject matter experts are already working with members of the relevant professional community to assess potential risks, privacy, and accuracy issues pertaining to the use of facial recognition technology.

Appendix IX: Comments from the Federal Bureau of Investigation

U.S. Department of Justice

Federal Bureau of Investigation

Washington, D. C. 20535-0001

March 19, 2021

Ms. Gretta L. Goodwin
Director
Homeland Security and Justice
Government Accountability Office
441 G Street, NW
Washington, DC 20548

Dear Ms. Goodwin:

Thank you for the opportunity to review and comment on the Government Accountability Office's (GAO's) draft report entitled FACIAL RECOGNITION TECHNOLOGY: Federal Law Enforcement Agencies Should Better Assess Privacy and Other Risks (GAO-21-243SU). The FBI agrees with the two recommendations addressed to the FBI Director. The FBI's planned actions to address the recommendations are addressed below:

Recommendation: The Director of the FBI should implement a mechanism to track what non-federal systems with facial recognition technology are used by employees to support investigative activities.

FBI's Response:

The FBI recognizes the importance of tracking the use of facial recognition technology by FBI employees, contractors, and Task Force Officers and currently follows numerous policies governing the use of such technology, including, but not limited to, the Domestic Investigations and Operations Guide (DIOG) and the Next Generation Identification Policy and Implementation Guide. The FBI will review and, if necessary, update existing policy to ensure this GAO recommendation is addressed.

Recommendation: The Director of the FBI should, after implementing a mechanism to track non-federal systems, assess the risks of using such systems, including privacy and accuracy related risks.

FBI's Response:

The FBI recognizes the importance of assessing privacy and accuracy related risks associated with FBI use of non-federal systems with facial recognition technology and is committed to assessing such risks in accordance with applicable law and policy. The FBI will review and, if necessary, update existing policy to ensure this GAO recommendation is addressed.

Again, thank you for the opportunity to comment on this report. We look forward to GAO closing the recommendations that the FBI has agreed to address.

Sincerely,

Darrin E. Jones
Executive Assistant Director
Science and Technology Branch

Appendix X: Comments from the United States Postal Service

UNITED STATES POSTAL SERVICE

March 5, 2021

Greta L. Goodwin
Director, Homeland Security and Justice
United States Government Accountability Office
441 G Street, NW
Washington, DC 20548-0001

SUBJECT: Draft report review of *Facial Technology Recognition: Federal Law Enforcement Agencies Should Better Assess Privacy and Other Risks* (GAO-21-243SU) – *March 2021*

Dear Ms. Goodwin:

Thank you for the opportunity to review and comment on the United States Government Accountability Office (GAO) draft report titled *Facial Technology Recognition: Federal Law Enforcement Agencies Should Better Assess Privacy and Other Risks.*" Our responses to the GAO's Recommendations for Executive Action are set forth below.

Specific Responses

With regard to your specific recommendations, we provide the following responses:

Recommendation 23

The Chief Postal Inspector of the U.S. Postal Inspection Service should implement a mechanism to track what non-federal systems with facial recognition technology are used by employees to support investigative activities.

Management's Response –

Management agrees in part with this recommendation. The U.S. Postal Inspection Service currently does track employee usage of agency-procured and provided non-federal systems with facial recognition technology. We will develop a mechanism to track other non-federal systems with facial recognition technology that our employees use to support investigative activities.

Expected completion date: September 30, 2021.

Recommendation 24

The Chief Postal Inspector of the U.S. Postal Inspection Service should, after implementing a mechanism to track non-federal systems, assess the risks of using such systems, including privacy and accuracy related risks.

475 L'Enfant Plaza SW
Washington, DC 20260-5665
www.usps.gov

- 2 -

Management's Response –

Management agrees with this recommendation. Risk assessments are continuous for any system(s), or usage of systems, that our agency procures, to include agency-procured and provided non-federal systems with facial recognition technology. Upon developing a mechanism to track other non-federal systems with facial recognition technology that our employees use to support investigative activities, we will assess the risks associated to such usage.

Expected completion date: March 31, 2022.

The Postal Service appreciates the opportunity to respond to GAO's draft report and Recommendations for Executive Action.

Sincerely,

Gary Barksdale
Chief Postal Inspector

cc: Sally K. Haring, Manager, Corporate Audit and Responses

Chapter 5

Facial Recognition Technology for Certain Traveler and Noncitizen Identity Verification*

Abigail F. Kolker and Kristin Finklea

Facial recognition technology (FRT), which uses algorithms to compare faces against a collection of known faces, is a biometric tool employed by law enforcement, including U.S. Customs and Border Protection (CBP) officials. FRT can be used to verify the identity of travelers and certain noncitizens awaiting entry.

FRT is used in two primary ways:

- identification searches, comparing features of a probe photo with those in a database of images to produce a gallery of potential matches ranked by similarity (not a single affirmative match), and
- identity verifications, comparing a probe photo against known photos to produce a binary, match/no match result.

Traveler Verification Service (TVS)

CBP employs FRT at many air, land, and sea ports of entry and preclearance locations, both domestically and abroad, and used it to process over 23 million travelers in FY2020. Specifically, CBP, in partnership with the Transportation Security Administration (TSA), uses FRT to verify travelers' identities. TVS

* This is an edited, reformatted and augmented version of the Congressional Research Service Publication No. IN11695, dated June 29, 2021.

In: Facial Recognition Technology
Editor: Mari F. Burke
ISBN: 979-8-88697-124-8

is a public-private partnership between the federal government and private airlines, airports, and cruise lines. It compares a traveler's live photograph (taken, for example, by a gate agent) to a gallery of preexisting photographs maintained by CBP. The composition of the galleries depends on the travel context. TVS provides a match or no match result within two seconds. In case of the latter result, the traveler's identity is checked manually by a CBP agent.

TVS Access through CBP One

In October 2020, CBP launched CBP One, a publicly available mobile application and portal for the public to interact with CBP. Its functionality differs depending on the type of user. For instance, cargo carriers can make appointments for the inspection of perishable items, travelers can apply for and view their I-94 forms, and international organizations (IOs) can verify the enrollment of individuals in certain programs—specifically, the Migrant Protection Protocols (MPP) program.

Migrant Protection ProtocolsProgram

MPP, or "Remain in Mexico," was broadly implemented under the Trump Administration in mid-2019 to address what DHS described as a "security and humanitarian crisis on the Southern border." MPP required many asylum seekers who arrived at the southern border to wait in Mexico while U.S. immigration courts process their cases, a departure from the previous policy of releasing some (particularly those in family groups) into the United States while they wait. The Biden Administration, after suspending enrollments in January 2021, ended the program and stopped new enrollments in June 2021. Approximately 68,000 individuals were enrolled in MPP and returned to Mexico. As of February 2021, there were approximately 25,000 MPP enrollees outside of the United States awaiting adjudication of their cases.

IOs, including the United Nations High Commission on Refugees (UNHCR) and the International Organization of Migration (IOM), provide humanitarian and legal assistance to MPP enrollees as they wait in Mexico. These organizations "identify MPP enrollees, verify their identity and MPP status, and transport them to a designated port of entry" at the date and time listed on their Notice to Appear document. CBP has partnerships with specified IOs, identified by the Department of State, to assist MPP enrollees. These organizations have access to the international organization functionality on CBP One.

IOs can use CBP One to verify that an individual is enrolled in MPP and has a pending immigration case. With the individual's consent, IO users upload a photograph of the individual to CBP One, which searches for a match against a pre-staged TVS photo gallery comprised of approximately 70,000 images of MPP enrollees saved in the Department of Homeland Security's (DHS's) Enforcement Integrated Database (EID). These images were collected during MPP enrollees' initial encounters with CBP. Just as TVS provides a match/no match response for travelers, it also does so for photographs that IOs enter in CBP One to verify the identity of MPP enrollees.

- If a match is confirmed, CBP sends the biographic information associated with the EID image (e.g., first name, last name, date of birth) to the Person Centric Query System (PCQS) to verify the individual has a pending immigration case. The IO user then receives one of four responses: (1) a green check mark indicating the individual is enrolled in MPP and has a pending immigration case; (2) a yellow bar indicating either the individual is enrolled in MPP but their case is closed or the individual is enrolled in MPP but CBP could not locate their immigration case (the IO user can then check the Department of Justice's Executive Office for Immigration Review website to confirm the case status); (3) a red "X" indicating CBP was unable to locate the individual's information; or (4) a system error message.
- If the submitted photograph results in a no match or if the individual elects not to provide a photograph, the IO user can submit the individuals' Alien Registration Number ("A-Number") instead. This will also query EID and PCQS and result in one of the four responses listed above. If CBP is still unable to confirm a match, it can submit the individual's biographic information (e.g., first name, last name, date of birth) to CBP One and repeat the same process.

Accuracy of TVS

One of the primary concerns over use of FRT has been its accuracy. While the reliability of FRT algorithms has improved, accuracy rates can vary based on demographic factors. An errant match (false positive) in TVS could pose a security risk, while an errant no match (false negative) could cause travel or immigration delays. CBP internal analysis estimates the false positive rate of TVS at .0103%. (It did not report the false negative rate.) Policymakers may exercise oversight over how CBP implements—including training for officials using FRT—and evaluates the accuracy of TVS and related systems as their use evolves.

Chapter 6

U.S. Capitol Attack and Law Enforcement Use of Facial Recognition Technology*

Kristin Finklea and Kelsey Y. Santamaria

On January 6, 2021, U.S. Capitol security was violently breached while Congress was in session to vote on the certification of the 2020 presidential election. Many participants in the attack reportedly intended to disrupt this process. Some clashed with law enforcement, leaving five dead and many more injured. Some have been charged with violations including crimes against persons and federal property. Some allegedly carried firearms and other weapons. Law enforcement also uncovered explosive devices near Capitol grounds.

Participants, media, and others documented the attack on the Capitol in photographs and video. Law enforcement, such as the Federal Bureau of Investigation (FBI), is seeking this digital content and other evidence to assist its investigation. Among the tools that law enforcement agencies may employ to identify potential criminal suspects depicted in digital content is facial recognition technology (FRT). While some have reported a possible uptick in law enforcement use of FRT following the Capitol attack, the extent to which FRT has been used to identify suspected rioters is unknown.

---

* This is an edited, reformatted and augmented version of Congressional Research Service Publication No. IN11614, dated February 24, 2021.

In: Facial Recognition Technology
Editor: Mari F. Burke
ISBN: 979-8-88697-124-8

Law Enforcement Use of FRT

FRT, which compares images of faces using facial geometry, is one biometric tool employed by law enforcement. FRT can be used to generate suspect leads, identify victims, sort faces in photos that are part of forensic evidence, and verify the identity of inmates being released from prison. FRT is generally used by law enforcement in *one-to-many searches* (comparing features of a *probe* photo with those in a database of images) to produce a gallery of potential suspects ranked by similarity—not a single affirmative match.

FBI Use of FRT

The FBI operates two FRT programs: (1) the Next Generation Identification–Interstate Photo System (NGI-IPS), largely supporting state and local law enforcement; and (2) the Facial Analysis, Comparison, and Evaluation (FACE) Services Unit, supporting FBI investigations.

- NGI-IPS contains criminal mugshots, and the system allows authorized law enforcement users to search probe photos of unknown persons against faces in the database for potential investigative leads.
- The FACE Services Unit searches probe photos against faces in NGI-IPS and other authorized federal and state facial recognition systems.
- A facial recognition search alone cannot provide law enforcement with a positive identification; the results must be manually reviewed and compared by an officer trained in facial comparison. The FBI prohibits law enforcement agencies from taking action (e.g., making an arrest) based solely on the results of a search in NGI-IPS.

There is a patchwork of FRT systems used by federal, state, and local law enforcement agencies around the country—including those investigating the Capitol attack. Each has its own database of faces against which probes may be compared, and law enforcement agencies may have differing policies governing FRT use. For instance, federal law enforcement agencies and authorized users of their FRT systems must generally follow established FRT policies and guidelines, which may not apply to state and local law enforcement agencies' FRT systems.

Federal law enforcement agencies have previously used FRT to identify individuals suspected of criminal activity associated with protests. Federal agencies investigating the Capitol attack have not confirmed specific use of FRT but have confirmed reviewing photo and video footage. State and local law enforcement, and individuals and companies, may also use FRT to help identify suspected Capitol rioters—tips that they may provide to law enforcement.

Concerns over the use of FRT include the accuracy or reliability of FRT systems, public notification regarding the use of FRT, and policies governing law enforcement agencies' use of the technology.

Concerns surfaced following reports that Clearview AI, a company that developed image-search technology used by law enforcement agencies around the country, had amassed a database of over 3 billion images against which probe photos could be compared. Some concerns have manifested in federal, state, and city efforts to prohibit or bound law enforcement agencies' use of FRT. Further, some companies producing facial recognition software have enacted barriers to law enforcement using their technologies.

Legislative Considerations

FRT can be a powerful investigative tool for law enforcement. But some observers have voiced concern about the technology's current and prospective use. While FRT's reliability has improved over time, the accuracy rates of FRT systems vary, particularly in identifying persons in certain demographic groups, leading some to express concern regarding possible misidentification. Some have also expressed concern that FRT—when paired with other tools and databases that contain millions of face images—could enable surveillance that encroaches on personal privacy and civil liberties. Others contend that these concerns are overstated and may not justify curtailing law enforcement's use of this investigative tool.

Several bills in the 116th Congress sought to regulate the use of FRT by the government or commercial entities, with most seeking to constrain its use. Several bills would also have sought to influence how state and local police use FRT by conditioning the receipt of criminal justice grants upon compliance with certain guidelines. Since the Capitol attack, there has been considerable discussion of whether police use of FRT to identify suspected participants will inform policymakers' assessment of whether to encourage or constrain the police use of FRT in the future.

Congress has not enacted legislation specifically addressing FRT. Outside the border security context, the most pertinent federal laws address the collection and storage of personal data by government agencies or commercial entities generally, though many such laws do not constrain information shared for law enforcement purposes. State and local regulation of FRT, in turn, varies considerably. While many state and local governments widely employ FRT, others expressly prohibit or limit its use. And although the Constitution provides baseline parameters for FRT's use by government actors, these considerations may be more relevant to assessments of the technology's use in a particular investigation than to FRT's general development and deployment.

Index

A

B

C

F

G

S

T

U

V